Elementary Kinetics of
Membrane Carrier Transport

K. D. NEAME & T. G. RICHARDS

DEPARTMENT OF PHYSIOLOGY
UNIVERSITY OF LIVERPOOL
LIVERPOOL, ENGLAND

Elementary Kinetics of Membrane Carrier Transport

BLACKWELL SCIENTIFIC PUBLICATIONS

OXFORD LONDON EDINBURGH MELBOURNE

ISBN 0 632 08100 7

First published 1972

Printed in Great Britain by
Adlard & Son Ltd
Bartholomew Press
Dorking, Surrey
and bound at
Kemp Hall Bindery
Oxford

317483

Contents

Preface

This book has been written principally for those with little or no previous experience in the kinetics of carrier transport, but those at a more advanced level should also benefit since it aims to provide understanding of a field in which experimental planning and sometimes the interpretation of experimental results are often questionable. We have tried to produce a rational framework within the limits of present knowledge which may help to avoid fundamental errors and upon which ideas may be developed.

The subject material has been mainly confined to the theory of the analysis of solute movement in terms of diffusion, adsorption, carrier transport and competition, without attempting to define a 'carrier' in terms other than kinetic. We have tried to give the theory support by referring to experimental work on the transfer of organic metabolites such as amino acids and simple sugars rather than inorganic ions such as potassium and sodium. (With the last two, the kinetics of transfer is made more complex by the particular relationships with the electrical charge on biological membranes.) The structure of membranes, the possible chemical interactions of transferred solute with membrane constituents, the nature of reactions with those constituents, the isolation of carriers, and the energetics of transfer are not discussed.

Where possible, illustrations have been taken from published reports, and some have been included rather for their suitability to illustrate particular points than as adequate models of technique.

The mathematics has been kept as simple as possible; the equations are almost all based on one or two simple expressions, and essentially the same basic symbols have been used throughout. Equations which may appear at first glance to be complex are usually only the result of algebraical manipulations of a more simple expression.

Acknowledgments

We would like to thank authors and publishers for their kind permission in allowing us to reproduce the following diagrams: Fig. 54 (G.Levi, R.Blasberg and A.Lajtha), from *Archives of Biochemistry and Biophysics*; Fig. 33 (L.R.Finch and F.J.R.Hird), Fig. 36 (F.Alvarado and R.K.Crane), Fig. 48 (H.N.Christensen) and Fig. 52 (J.A.Jacquez) from *Biochimica et Biophysica Acta*; Fig. 26 (D.M.Matthews, I.L.Craft, D.M.Geddes, I.J.Wise and C.W.Hyde) from *Clinical Science*; Fig. 17 (C.G.Winter and H.N.Christensen), Figs. 21 & 40 (H.N.Christensen and M.Liang) and Fig. 49 (H.N.Christensen, D.L.Oxender, M.Liang and K.A.Vatz) from the *Journal of Biological Chemistry*; Fig. 12 (T.G.Bidder), Fig. 20 (L.L.Iversen and M.J.Neale) and Fig. 53 (D.D.Wheeler and L.L.Boyarsky) from the *Journal of Neurochemistry*; and Fig. 42 (E.Epstein, D.W.Rains and O.E.Elzam) from the *Proceedings of the National Academy of Sciences*.

Symbols and Equations

A. SYMBOLS

b Dissociation constant for adsorption, page 21

i   Inhibitor, page 57

K_D Diffusion constant, page 6

k_i Inhibitor constant for adsorption, page 58

K_i Inhibitor constant for carrier transport, page 60

K_m Michaelis constant for carrier transport, page 24

K_a Apparent K_m; the concentration of substrate in a system following Michaelis–Menten kinetics, at which half the maximal rate of transport is developed in the presence of a fixed concentration of competitive inhibitor, page 60

K_p Pseudo-K_m; the concentration of substrate at which half the maximal rate of transport is developed by a carrier system which as a whole does not follow Michaelis–Menten kinetics, page 89

K_s Equilibrium constant for adsorption, page 21

m Mass of substrate in fluid bathing adsorption surface, page 19

Q Quantity or mass of substrate on adsorption surface or carrier of limited capacity, page 20

Q_{max} Maximal quantity or mass of substrate which could theoretically occupy adsorption surface or carrier of limited capacity, page 20

r That fraction of the total number of molecules striking an adsorption surface in unit time which becomes attached, page 20

S   Substrate, page 5

V Overall or net rate of transfer, page 5

v Rate of transfer in one direction only, page 4

v_{max} Maximal rate of transport by carrier in one direction only, page 25

α Association constant, page 20

β 'Backwards' dissociation constant for carrier, page 24

γ 'Forwards' dissociation constant for carrier, page 24

θ Fraction of adsorption surface occupied by substrate, page 18

[ ]   Concentration of substance designated, page 5

'   Orientation with respect to second (or intracellular) phase as opposed to first (or extracellular) phase, page 5

B. BASIC EQUATIONS

Equation
No.

One-way transfer:

$$\text{(a)} \quad v = K_D.[S] \qquad\qquad \text{(b)} \quad v = \frac{v_{max}.[S]}{[S]+K_m}$$

(9), (27)

(diffusion)   (carrier transport)

Overall or net transfer:

$$V\,(=v-v') = \left(\frac{v_{max}.[S]}{[S]+K_m}+K_D.[S]\right) - \left(\frac{v_{max}'.[S']}{[S']+K_m'}+K_D.[S']\right)$$

(36)

('inward' transfer)   ('outward' transfer)

One-way carrier transport of substrate in the
presence of competitive inhibitor:

$$v = \frac{v_{max}.[S]}{[S]+K_m\left(1+\dfrac{[i]}{K_i}\right)}$$

(69)

Exponential change of concentration with time:

$$[S]_t = A.e^{-kt}+B$$

(12)

Chapter 1

Introduction

The biological cell is a self-contained structure in which innumerable metabolic processes occur. These may involve the breakdown and production of organic materials, and require a continuous removal and replacement of metabolites in order to maintain the integrity of the cell. The disposition of these materials is in many cases only achieved by movement across a membrane, whether it be the membrane boundary of the cell as a whole or that of an intracellular organelle, and it is with the observed characteristics of this movement that we are principally concerned.

The evidence we have upon which to base our views on the nature and ultrastructure of cell membranes will not be discussed here, but the reader is referred to other works which deal with these matters [1, 27, 28, 29, 44, 45, 90, 114a, b, 129, 130, 138]. At the present time there is no firm agreement about the organization of the living membrane, whose structure may vary from cell to cell, and with the state of activity of the membrane. The membrane itself may vary from a simple envelope to a structure with numerous and tortuous invaginations lining channels and spaces which run through the substance of the cell. For the purpose of kinetic analysis we must, however, make certain simplifying assumptions. The cell or tissue is treated as if it were effectively a uniform compartment bounded by a uniform membrane, that is, by a simple envelope. The solute transferred across is assumed also to be in simple solution whichever side of the membrane it is on, and the quantity which moves across is calculated from the change in concentration on one or both sides of such a membrane.

It is possible that a solute whose movement is measured may not be in simple watery solution within a cell, even though it may be extracted as if it were. There is evidence to suggest that the organization of water molecules in a cell may be more crystalline than fluid [55]; in addition, water molecules may be 'bound' to protein molecules in the cell, or their density may not be the same in all parts of the cell. Variations of this sort would affect the movement of molecules in solution as a result of varying solubility. Any kinetic description based upon experimental observations will take these into account without necessarily describing them in detail.

Transfer across biological membranes has been postulated as being of two main types, 'diffusion' and 'carrier transport'. *Diffusion* involves movement entirely under the control of physical forces and without direct interaction with tissue components; the

1

rate of transfer is directly proportional to the concentration of the solute, and is affected by factors which are usually constant for a particular set of experiments (e.g. temperature, viscosity of solvent, size of diffusing molecule, and nature of the membrane) [45]. *Carrier transport* is a term used to describe transfer which is at a higher rate than would be attributable to diffusion alone and which shows many of the properties of enzyme kinetics [159]. For example, at lower concentrations it can appear to resemble diffusion, but at higher concentrations it shows evidence of saturation, the rate of transfer approaching a maximum value in spite of a progressive increase in the concentration of solute presented to the membrane [21, 153, 154]. The term 'carrier' is simply a way of explaining such phenomena, and the appearance of saturation is attributed to the occupation of most of the limited number of sites in the membrane to which it is believed that the solute must become attached before being transferred across. No structure in the membrane has yet been definitely identified as a carrier, but a conformational change which has the same effect as a structure moving across the thickness of the membrane would explain much of the experimental evidence.

Many kinetic and mathematical models have been proposed [7, 24, 29, 42, 69], most of which, but not all [122], are based on this simple concept. Although grossly oversimplified, it suffices for the development of a theory of transfer kinetics, and we use the word 'carrier' in the absence of a better alternative.

Unlike an enzyme, a carrier, whatever may be its form, cannot as yet be isolated and still demonstrate the property of transfer. If it lies on the boundary between two compartments, isolation implies its removal from such a boundary, hence the disappearance of demonstrable transfer. Identification of carrier has been attempted in the erythrocyte membrane by a study of carrier energy relationships [7], and in the intestine and kidney by the observation of adsorption of solute at the cell membrane [61, 117].

Carrier transport is generally divided into two types. In one type solute usually moves down a concentration (or electrochemical) gradient which, assuming no other physical forces to be present, would eventually lead to equal concentrations of the solute on either side of the membrane; in this respect it resembles diffusion, but the evidence suggests that a carrier is involved in transfer. This type we shall refer to as *equalizing transport*. In the other type, transfer can in time reach an equilibrium associated with a concentration of solute on one (usually the intracellular) side of the membrane which is higher than that on the other side, and the transfer requires energy from cell metabolism. This type we shall call *concentrative transport*.

Chapter 2

Diffusion

Diffusion has been defined as the migration of molecules from a region of higher concentration to one of lower concentration as a result of their random motion [45], and in accord with the second law of thermodynamics. This law [57] describes a system in which energy will spontaneously pass from a region of higher energy to a region of lower energy, so that entropy, which is a measure of 'randomness' or 'state of chaos' in the total system, is increasing.

Many life-processes, and the organizations upon which they depend, can produce a local decrease in the state of entropy by the use of energy derived from external sources. For example, barriers in the form of membranes, built and maintained by the use of metabolic energy, may retard progress towards a state of greater entropy by reducing the rate of movement of molecules.

A good description of the kinetics of diffusion (see also Riggs [128]) has been written by Hartley & Crank [59]. These authors take as their example the diffusion of iodine molecules in a solvent; we give here a modified version of their account.

Let us consider a container in which there are two solutions of identical molecules, both dilute, but one more dilute than the other. The two solutions are in contact along an initially sharply defined boundary and they are not separated by any membrane. All the molecules are in random motion as a result of thermal energy. There is no preferred direction of movement of any one molecule, since in a dilute solution each molecule is virtually unaffected by the presence of others. Thus they move sometimes towards a region of higher concentration and sometimes towards a region of lower concentration. Consider now two thin elements of equal volume, one on either side of the boundary. This is illustrated in greatly simplified form in Fig. 1, in which only a few molecules adjacent to the boundary are shown. One element of volume (A) contains molecules at the higher concentration and the other (B) at the lower concentration. As a result of their random movement, a certain fraction of the molecules from element A will move across to element B and vice versa. In Fig. 1 (upper sketch) there are eight molecules on one side of the boundary and four on the other, and during a specified but short interval of time, one quarter of the molecules from each side move across to the other side of the boundary. This situation is shown in the middle sketch of Fig. 1. The ratio of molecules on the two sides in the figure has now been reduced from 8 : 4 to 7 : 5.

The process can be repeated for a subsequent interval of time, starting with the new (7 : 5) relationship. If it were repeated a very large number of times, the ratio would approach unity and the number of molecules on each side would be nearly equal; eventually equilibrium would be reached (Fig. 1, lower sketch). The changes in concentration are exponential; that is to say, over similar intervals of time the number present on each side will alter by the same proportion, comparable to the progress of radioactive decay. Although there is no preferred direction of movement of any particular molecule at any time, an equilibration between the two groups of molecules by exchange will, nevertheless, be reached in due course.

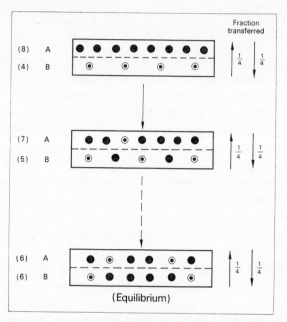

Fig. 1. Diffusion between two compartments: a simplified version. *Upper:* the upper compartment (A) initially contains eight molecules, the lower compartment (B) four molecules, and in any unit of time one-quarter of each group of molecules move across the boundary into the other compartment as a result of thermal agitation. *Middle:* after one unit of time two molecules have moved from the upper to the lower compartment and one from the lower to the upper, so that there has been a net transfer of one molecule from the upper to the lower compartment. *Lower:* eventually equilibrium exists, and although one-quarter of the molecules in each compartment still move across the boundary in unit time, there is now no net transfer. (Movements are normally random and in any direction, and consistent proportionate changes will only occur with large numbers of molecules.)

The proportion of molecules which move across a boundary in a given system is constant for a particular solute. Since the number of molecules available is proportional to their concentration, the number of molecules moving across unit area in unit time is also proportional to that concentration.

This rate of mass transfer, often referred to as the 'velocity' of transfer may be expressed by the symbol v (mass/unit area of boundary in unit time) and related to the concentration of molecules, expressed by the symbol [S] ('S' being derived from the word 'substrate' used in later discussions, which refers to a solute whose molecules are able to combine reversibly with adsorption sites), thus:

$$v \propto [S] \tag{1}$$

(In enzyme kinetics it has become conventional for 'velocity' or 'rate' to apply to a change of structure; in carrier kinetics it refers to a change of position, but without any linear connotation.)

The rate of transfer in the opposite direction (v') can clearly be related to the concentration on the other side of the boundary ([S']), so that:

$$v' \propto [S'] \tag{2}$$

The overall or net rate of transfer (V) at any moment of time is the difference between the two separate rates, and hence is proportional to the difference between the concentrations on either side of the boundary:

$$V = (v - v'), \quad \text{or} \quad V \propto ([S] - [S']), \tag{3}$$

$([S] - [S'])$ representing the energy difference between the two sides of the boundary.

Diffusing molecules are thus said to 'pass down a concentration gradient'. In fact, it is the *net* mass movement which may be so regarded. At all times movement takes place in each direction, and any designated molecule may participate in either of these movements, from one 'compartment' to another, several times during any period of observation. One cannot therefore describe diffusion into a cell without making a statement about diffusion out of the cell.

The boundary has been treated here as if it had no thickness. In practice, molecules move across boundaries of finite thickness, l, across which there is a progressive change in concentration gradient. This could be expressed as a difference in the concentration of molecules per unit of membrane thickness, i.e.

$$\frac{[S] - [S']}{l},$$

so that the relationship with the rate of transfer becomes:

$$v \propto \frac{[S] - [S']}{l}. \tag{4}$$

(This is essentially Fick's Law, which says that the rate of diffusion of a dissolved substance along a column of fluid is proportional to the concentration gradient [55a].)

In biological terms the 'boundary' is a membrane across whose thickness there may be a concentration difference, and through which there is assumed, for this purpose, to be a uniform gradient.

The concentration of solute in the fluids on either side of a membrane may be assumed to be the same at all points if the rate of diffusion within the solvent is much faster than through the membrane [45] or if the fluids are kept well mixed. This requirement is easily met for a medium which surrounds tissue slices. It may well be, however,

B

that local differences of intracellular concentration do exist in reality, although the present state of knowledge does not allow us to describe the kinetics of such a system in any specific case. For this reason biological systems are, in many cases, treated as if solute exists in a single well-mixed compartment.

Numerous factors will alter the rate of transfer of molecules across a membrane; for example,

(i) Temperature (which will alter the rate of random movement of molecules).

(ii) 'Viscosity' between solute and the solvent through which it diffuses (which will affect the freedom of movement of solute molecules).

(iii) The size and shape of solute molecules in relation to the organization of 'immobile' constituents of a membrane, i.e. 'porosity'.

(iv) Electrical charges.

(v) Solvent flow in the membrane either with or against the direction of net solute transfer, i.e. 'solvent drag'. This is, in effect, a variant of (ii) above.

(vi) Relative solubility of solute molecules in the solvent on either side of the membrane and in the membrane itself.

(vii) The surface area of membrane accessible to solute.

In a specified experimental system these factors can be assumed to be constant. They can all therefore be incorporated into a single constant, the *diffusion coefficient*, designated by the symbol D [45, 57], and the general relationship can then be expressed as an equation:

$$V = \frac{D\,([S]-[S'])}{l} \tag{5}$$

which can also be written:

$$D = \frac{V.l}{[S]-[S']} \tag{6}$$

showing that D represents the 'rate of movement (i.e. mass/unit area of membrane in unit time) × unit of membrane thickness/unit of concentration difference'. Its dimensions are therefore

$$\frac{mass}{area \times time} \times length \times \frac{volume}{mass}$$

$$\left(e.g.\ \frac{moles}{cm^2 \times min} \times cm \times \frac{cm^3}{moles}\right)$$

which simplifies to

$$\frac{area}{unit\ time}$$

(e.g. cm^2/min).

In most biological systems it is not easy to measure the thickness of a membrane, which is therefore usually incorporated as part of a single constant, the *permeability constant*, which is represented by D/l. In many biological investigations, however, the conformation of a membrane is not easily defined, and so a constant which is expressed in terms of the tissue rather than of its membrane, and which is called the *diffusion constant*, i.e. K_D, is more often used.

The equation is now written:

$$V = K_D ([S] - [S'])$$ (7)

or alternatively:

$$K_D = \frac{V}{[S] - [S']}$$ (8)

The diffusion constant is fundamentally a measure of the rate at which molecules of a particular solute penetrate a defined region of membrane or a defined mass of tissue. It describes a relationship between a solute, a solvent and the membrane and is, therefore, only partially descriptive of any one of them. The dimensions will depend on the mode of expression, and in comparing work from different sources it may be necessary to convert values expressed in one way in order to compare them with those expressed in another. In some cases, as with the erythrocyte, the area of membrane can be reasonably accurately measured. K_D can then be equated with the permeability constant, and will be measured as '(rate of movement/unit area of surface) × unit of concentration difference'

e.g. $\dfrac{\text{moles}}{\text{cm}^2 \times \text{min}} \times \dfrac{\text{cm}^3}{\text{moles}}$

which simplifies to 'distance/unit time' (e.g. cm/min). When the area of membrane cannot be measured, as with tissue slices, 'area' can be replaced by 'volume' or 'mass' of tissue, so that 'cm^2' would be replaced in the expression by 'cm^3' or even by 'grams'. In the former case the constant will simplify to 'per unit time' (e.g./min, or min^{-1}) [114]

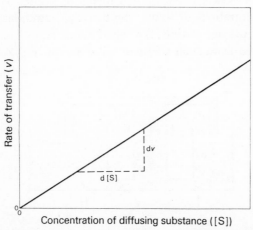

Concentration of diffusing substance ([S])

Fig. 2. Relationship between the one-way rate (v) of diffusion of molecules across a boundary and the concentration of those molecules in the compartment from which they are diffusing ([S]). Rate is proportional to concentration, as expressed by the equation $v = K_D.[S]$. K_D is the diffusion constant, and represents the slope of the line, i.e. $dv/d[S]$, where 'd' indicates a small quantity. Net movement (V) and the difference between the concentrations on each side of the boundary ([S] − [S']) may replace v and [S] respectively on ordinate and abscissa; the corresponding equation would then be $V = K_D ([S] - [S'])$, and K_D would then be equal to $dV/d([S] - [S'])$.

and in the latter to 'volume/(unit mass × unit time)' (e.g. cm^3/g tissue.min) or some similar arrangement [144]. We are here assuming a constant relationship between membrane area and tissue mass.

It is not assumed that all parts of the area of membrane involved in the observed action behave identically at all times or at any time, but it is assumed that such an asymmetry applies equally to substrate on either side of the membrane, so that all the factors which go to make up K_D will apply equally to movement in either direction. K_D can therefore alternatively be applied to the two opposite one-way components (v and v') which make up net movement. These two movements can then be expressed as separate equations:

$$v = K_D.[S] \text{ and } v' = K_D.[S'], \qquad (9), (10)$$

net movement then again being expressed as:

$$V = (v - v') = K_D([S] - [S']). \qquad (11)$$

A plot of rate of movement against concentration or concentration difference results in a straight line which passes through the origin (Fig. 2). Either type of equation (i.e. for one-way movement or for net movement) corresponds to the simple algebraical expression for a straight line, $y = ax$; K_D then corresponds to a and on the plot would be represented by the slope of the line.

Diffusion and equilibrium in a two-compartment system

When diffusion occurs between two compartments of limited volume which initially contain different concentrations of solute, the two concentrations approach each other exponentially with the passage of time. An experimental system of this type is shown in Fig. 3, and data derived from such a system are shown in Fig. 4. Since the volume of

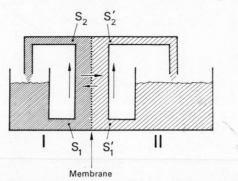

Membrane

Fig. 3. A two-compartment experimental system. In each compartment a pump (not shown) circulates fluid past a dialysing membrane (*vertical interrupted line*) and then back to a reservoir. The fluid in the left-hand compartment (I), which initially contains solute, is half the volume of that in the right-hand compartment (II), which initially contains no solute. Samples are taken in both compartments at the points shown, that is, at S_1 and S_2 in Compartment I and at S_1' and S_2' in Compartment II. A plot of data from an experimental system of this type is shown in Fig. 4.

Compartment II is twice that of Compartment I and the initial concentration in Compartment II is zero, the concentration in both compartments at equilibrium is in this case one-third of the initial concentration in Compartment I, as shown in Fig. 4 by the horizontal interrupted line marked $[S]_{eq}$.

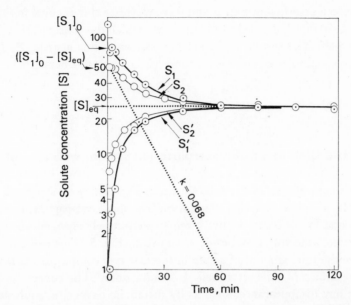

Fig. 4. Experimental data of diffusion in a two-compartment system of the type shown in Fig. 3. The data on each plotted line represent the concentrations of solute in samples taken at the four points shown in Fig. 3. Thus the *heavy continuous lines* represent concentrations of solute before the circulating fluid has reached the membrane and the *light continuous lines* those after it has left the membrane. The *horizontal interrupted line* represents the concentration at equilibrium, which is one-third of the initial concentration in Compartment I of Fig 3, reflecting the 1 : 2 ratio of the fluid volumes. The *diagonal interrupted line* is the exponential component of the *upper line*, S_1, and is obtained by subtracting the concentration at equilibrium ($[S]_{eq}$) from the values of that line (i.e. $= [S_1] - [S]_{eq}$). The concentration, $[S_1]$, at any time, t, may then be found from the exponential equation, $[S_1]_t = ([S_1]_0 - [S]_{eq}).\exp(-kt) + [S]_{eq}$ (eqn. 13). In this case, $[S_1]_0 = 59 + 25 = 84$ (the sum of the intersections of the two interrupted lines with the ordinate), $[S]_{eq} = 25$, and $k = (0.693/10.2) = 0.068$. Units are arbitrary; semi-log scale. (Obtained from experiments on the diffusion of K^+ ions from a solution of KCl; unpublished data.)

If the value of $[S]_{eq}$ is subtracted from any desired number of values in the upper curve of Fig. 4, a semilogarithmic plot of the resulting values gives a straight line (shown by the diagonal interrupted line in Fig. 4). This is the equivalent of an exponential curve on a linear plot having a base-line of zero, the experimental values (upper curve of Fig. 4) thus being the sum of this and the constant, equilibrium value $[S]_{eq}$. The experimental values can therefore be described by the exponential equation:

$$S_{(t)} = A.\exp(-kt) + B \qquad (12)$$

which will here have the form:

$$[S_1]_t = ([S_1]_0 - [S]_{eq}) \exp(-kt) + [S]_{eq} \tag{13}$$

where $[S_1]_t$, $[S_1]_0$ and $[S]_{eq}$ represent the concentration of solute in the fluid at S_1 (Fig. 3) at time t, at time zero, and at equilibrium respectively; the constant, k, is the slope of the exponential line in Fig. 4, and its numerical value is equal to the exponential decay constant, i.e. 0.693, divided by the time over which the concentration ($[S_1]_0 - [S]_{eq}$) is reduced by half. The second term, i.e. the horizontal line at $[S]_{eq}$, is $B.\exp(-ct)$, where the slope, represented by the constant 'c', is zero, and hence

$$B.\exp(-ct) = B \ (=[S]_{eq}),$$

since $\exp(0) = 1.0$.

Diffusion in a three-compartment system with run-off

A two-compartment system of the type shown in Fig. 3 may be separated from a third compartment by a second dialysing membrane. The third compartment may be so large that the concentration of solute within it remains effectively zero, and no solute returns through the interconnecting membrane, as shown in Fig. 5. This is often referred to as 'run-off', since the total amount of solute in the first two compartments is progressively reduced in the course of time, eventually becoming zero. (The concept of run-off need not be literal; any disappearance, such as by metabolic change, is equivalent.)

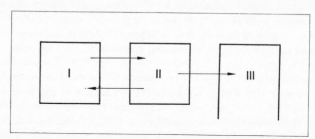

Fig. 5. A three-compartment system. Solute is initially only present in Compartment I. Movement takes place in the direction shown by the arrows, with 'run-off' in Compartment III.

An experimental system of this type if shown in Fig. 6, where the third compartment consists of a flow of solute-free fluid past the membrane. If we start with solute present only in the left-hand compartment (I), there is net transfer of solute to the middle compartment (II) and thence to the run-off compartment (III). Equilibrium is not achieved, since Compartments I and II are being continuously depleted, while net transfer demands that the concentration in Compartment I is greater than that in II at all times.

Fig. 7 shows a semilogarithmic plot of concentration against time in a system such as this. The equilibrium component shown in Fig. 4 is replaced by a second exponential component. The upper curve, which represents the concentration of solute at S_1 in

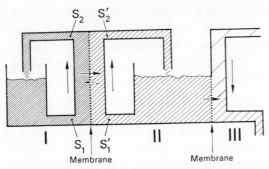

Fig. 6. A three-compartment experimental system. In each of Compartments I and II a pump (not shown) circulates fluid past a dialysing membrane, returning it to a reservoir. A third compartment (III) is separated from Compartment II by a second dialysing membrane across which is a continuous flow of fluid running to waste. (It is assumed that this flow is so fast that solute molecules which enter it from Compartment II have negligible chance of returning to Compartment II.) The volume of fluid in Compartment II is twice that in Compartment I, and solute is initially present only in Compartment I. Samples are taken at points S_1, S_2, S_1' and S_2'. A plot of data from an experimental system of this type is shown in Fig. 7.

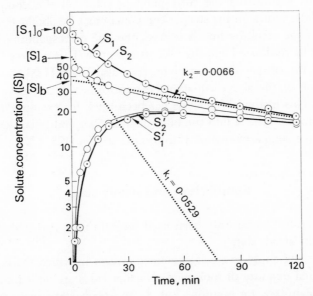

Fig. 7. Experimental data of diffusion in a three-compartment system of the type shown in Fig. 6. The data on each plotted line represent the concentrations of solute in samples taken at the four points, S_1, S_2, S_1' and S_2' shown in Fig. 6. The two *interrupted lines* represent the two exponential components whose sum at any time is equal to the observed concentration at S_1 at that time (*upper curve*). The two lines cross the ordinate at $[S_a] = 59$ and $[S_b] = 37$ respectively; $k_1 = 0.0529$ and $k_2 = 0.0066$. The concentration $[S_1]$ at any time, t, may be found from the double exponential equation $[S_1]_t = [S_a] \cdot \exp(-k_1 t) + [S_b] \cdot \exp(-k_2 t)$. Units are arbitrary; semi-log scale. (Obtained from experiments on the diffusion of K^+ ions from a solution of KCl; unpublished data.)

Fig. 6 consists of the sum of two exponential components (shown by interrupted lines); the steeper one reflects primarily (but not entirely) the relationships between Compartments I and II, and the shallower one is similarly related to movement between Compartments II and III. The upper curve ($[S_1]$) can be described by an equation representing the sum of two exponential components, thus:

$$[S_1]_t = [S_a].\exp(-k_a.t) + [S_b].\exp(-k_b.t) \tag{14}$$

where $[S_1]_t$ represents the concentration of solute in the fluid at point S_1 at time t, $[S_a]$ and $[S_b]$ the concentrations represented by the intersection of the two plotted exponential components with the ordinate, and k_a and k_b the slopes of those two components, determined as described above for a two-compartmental system. The expression $[S_b].\exp(-k_b.t)$ is an exponential expression replacing the fixed value $[S]_{eq}$ of equation (13).

The two exponential components can be resolved as follows. The points at times less than about 60 min represent the sum of relatively large values of both components. For times greater than this the contribution by the steeper component has become so small that the observed values are almost entirely derived from the shallower component. A straight line is drawn close to those observed values and is extrapolated back to the ordinate; it gives rough experimental values for this component over the whole experimental period. By subtracting the corresponding values of this component from the experimental values, values for the steeper component may be obtained, and a line drawn through them. If the resulting line is curved, the line corresponding to the shallower component must be readjusted and the process repeated. This process of approximation is repeated until the steeper line is straight. (See also Riggs [128].)

The derivation of the above equation (14) and other details have been described elsewhere [127]. This method of analysis has been used, for example, to determine the loss of solute from the plasma into the bile; plasma, liver and biliary system were then treated as the three compartments of a diffusional system [125, 127].

Equilibrium and steady state

The term 'equilibrium' has already been used several times, and it should not be confused with the term 'steady state'.

'Equilibrium' may apply to a self-contained system in which there is no change in the identity or total quantity of molecules. No molecules are added and none are lost from the total system (see, for example, Fig. 3). In 'steady state' there is also no change in the total number of molecules in a system as a whole, but there is a 'turnover' of molecules. These are either effectively or in reality lost to the exterior and replaced by the addition of fresh molecules at the same rate, so that there is replacement without a change in the quantity in the system at any time. For example, steady state could exist across Compartments I and II of Fig. 6 if solute were added to Compartment I at the rate at which it was being lost to Compartment III. The two states are summarized in Fig. 8.

The description of equilibrium and steady state have here been confined to a two-compartmental system, but these distinctions apply also to other types of multi-compartment system. When transfer by means of 'carrier' systems is described, the same cryteria will apply. It is important to appreciate that one of the 'compartments' above could be replaced equally well by an adsorption surface without in any way affecting the parameters so that equilibrium in such a case would be achieved when the number of molecules leaving the surface in unit time would be equal to the number being adsorbed.

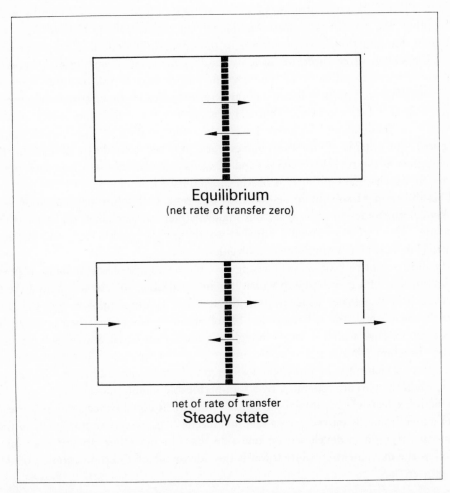

Fig. 8. Equilibrium and steady state. *Upper*: in *equilibrium* the rates of transfer in opposite directions are equal and the net rate is zero. *Lower*: in *steady state* the rates in opposite directions across the boundary are not equal so that there is a net rate in one direction, but the total number of molecules in the system remains constant.

Diffusion of solvent; osmosis

Solvent as well as solute will diffuse across a membrane when concentrations are different on the two sides. Solvent will diffuse down its own concentration gradient, and in general will move in a direction opposite to that of the solute. A concentration is the ratio of the number of molecules (or mass) of one species to the number of molecules (or mass) of a second species in solution with it. It follows that the larger the number of solute molecules in unit volume, the smaller is the number of solvent molecules in the same volume of solution. Thus if the concentration of solute is higher on one side of a membrane than on the other, the reverse is the case for the solvent concentration. Hence, if the solute diffuses in one direction, solvent diffuses in the opposite direction. This can be shown by the immersion of erythrocytes in a suspending medium consisting of a hypertonic solution of hexose in saline [154]. The erythrocytes rapidly shrink owing to movement of fluid out of the cells (where it is at a higher concentration) into the suspending medium (where it is at a lower concentration). Hexose enters the cells more slowly down its concentration gradient and its entry lowers the concentration of the fluid in the cells and raises that outside. Fluid then moves back into the cells down its concentration gradient (now in the reverse direction) and the cells consequently return to their original volume, at which time equilibrium has been established.

This diffusion of solvent in response to differences in solute concentration is the well-known phenomenon of osmosis, but the principles concerned are the same as those for solute. The forces eventually equalize so that, although there is still molecular 'exchange' there is no net migration of molecules.

The diffusion of solvent causes a change in volume and pressure. If solute is present only on one side of a membrane to which it is impermeable, solvent will migrate to that side since it is at a lower concentration there. Since solute cannot in this case move across the membrane, the net migration of solvent will only cease when opposed by a hydrostatic pressure which is large enough to produce an equal flow of solvent in the opposite direction. This is the 'osmotic pressure' of that solute and has been defined as 'the pressure which must be put upon a solution to keep it in equilibrium with the pure solvent when the two are separated by a semipermeable membrane' [151].

A similar effect will be obtained if solute, initially at equal concentrations on either side of a membrane, is moved across that membrane (e.g. by carrier transport) so that a larger concentration is developed on one side than on the other. In this case solvent will follow the movement of solute towards the side on which the concentration of solute is the greater.

Solvent drag

Movement of solvent through a membrane can also affect the movement of solute by a mechanism known as 'solvent drag'. Solute molecules crossing a membrane are in close association with solvent molecules; it there are no electrochemical interactions with the membrane, they will be entirely dependent on the solvent for their position in space.

If there are no other kinds of movement, solute molecules will follow the same overall movement as solvent molecules. This is similar to the movement of a balloon which is stationary in relation to the air mass containing it, and which as a result moves relative to the ground at the same rate as the air mass.

If solute is moving through a solvent, its movement relative to a membrane will then be a resultant of two movements: (i) the movement of solvent relative to the membrane, and (ii) the movement of solute relative to the solvent. The principle is similar to that of an aeroplane which moves through an air mass at a speed which depends only on the effect of its propeller on the air immediately surrounding it (air speed); at the same time, the air mass is moving as a whole over the ground (wind speed). The rate and direction of movement of the air relative to the ground (ground speed) is then a resultant of these two types of movement. The air speed is comparable with the movement of solute molecules relative to solvent, the wind speed with the movement of solvent through the membrane, and the ground speed with the movement of solute molecules themselves relative to the membrane. 'Solvent drag' refers to this effect which the movement of solvent has on the movement of solute molecules contained within it.

Chapter 3

Carrier Transport

Although diffusion is a fundamental mode of movement common to biological and non-biological systems, it cannot account for all types of movement across tissue membranes. With diffusion the rate of net transfer of solute is at all times proportional to the concentration difference across the surface through which diffusion takes place. In many biological tissues, however, it has been found that the rate of transfer across a membrane does not increase in proportion to the concentration of solute, but gradually approaches a limit at higher concentrations. At one time this was attributed to a possible change in the permeability of the cell membrane [71], but further investigation led to the view that a mechanism other than, or in addition to, diffusion was responsible. It has also been found that some metabolically-inert solutes may interfere with the transfer of other structurally-related solutes, apparently without chemical interaction between them. Behaviour such as this is now generally interpreted in terms of a 'carrier' function of some constituent of the cell membrane. A limited number of carriers are assumed to transfer solute across the membrane. When a site on the carrier is available to and can accept one of several solutes present at the same time, 'rivalry' between the solutes ensues, which is referred to as 'competition'.

Experimental evidence suggests that the transport of solute across a membrane by carrier may result in two conditions of equilibrium; in one, the concentrations of solute on each side of the membrane are the same at equilibrium, and this we shall call *equalizing transport*; in the other, they are different, and this we shall call *concentrative transport*. (In both cases, of course, net transfer at equilibrium is, by definition, zero.)

Equalizing transport has been given other names, such as *equalizing selective transport* [98], *facilitated transfer* [42, 45, 154], *facilitated diffusion* [29, 45, 159] and *assisted diffusion*. The use of the word 'diffusion' may be misleading since this type of transport behaves like diffusion only in restricted circumstances, although, of course, its presence does not preclude the co-existence of diffusion.

Concentrative transport has also been called *uphill transport* [29, 159] and *active transport* [42, 121, 159], but it is difficult to devise a satisfactory name [42]. A carrier potentially able to concentrate solute against a chemical gradient should not be excluded from this category when moving a solute down a gradient, nor should a carrier be included which can move solute only temporarily against a gradient (see Chapter 7, p. 95).

16

The simplest of the two processes is equalizing transport, and some mention of the data in support of its existence will now be given. A well-known example is the transport of certain sugars into the erythrocyte. It has been established that the proportional rate of transfer of these sugars is less at higher concentrations than at lower concentrations [71, 85]; this is sometimes described as a decrease in the value of an apparent diffusion or permeability constant as the sugar concentration is raised [154]. (The entry of other sugars, however, shows diffusion characteristics only [78].) This is not the only way in which the entry of these sugars into the erythrocyte appears to differ from diffusion [79, 154]; for example, sugars which are isomeric can have different rates of penetration [72, 76, 155] and sugars can interfere with each other's penetration according to their structure [76, 78]. There is no evidence, however, that at equilibrium the erythrocyte maintains a higher concentration of a sugar inside the cell than outside, and hence in this respect the process resembles diffusion.

The explanation of these observations has now been accepted in terms of a 'carrier' in the cell membrane. The membrane behaves as if it had a limited number of adsorption sites at each surface to which the sugar molecules become attached and which are then transferred to the opposite surface where the molecules become detached. Although such a carrier has never been directly observed, experimental data fit the kinetics which such a carrier model suggests [6, 154]. Attempts have been made to locate the position [61, 117] and to measure the concentration of such a carrier [47, 48]. There is also evidence that there are proteins with adsorptive capacity located in the cell membranes of bacteria [3a, 114a, 114b, 119a], which can be displaced into a suspending medium by 'osmotic shock' (exposure of the organism to water) [3a, 119a]. The evidence suggests that, while these proteins may be part of a carrier system, they are not themselves the agents which alone produce the translocation of solutes. For example, it has been found that in the membrane of mutant bacteria unable to transport sulphate there is a specific protein which is able to adsorb sulphate; this protein may be released into a surrounding medium by exposure to osmotic shock [114a, 114b]. Similarly, osmotic shock removes a galactose-adsorbing protein from other bacteria whose membrane is able to transport galactose, and at the same time causes loss of much of this transport capacity; the capacity may be partially restored by incubating the bacteria with dialysed 'shock fluid' (the water used for osmotic shock) [3a]. There is other evidence in this field (to be discussed briefly on p. 36) which suggests that the proteins referred to above are unlikely to be solely responsible for transport, but they may play an important role at some stage in the process.

A number of models have been evolved to describe qualitatively the behaviour of carriers. For example, there may be a 'shuttle service' between the two surfaces of the membrane [148], there may be a series of fixed attachment sites lining pores through the cell membrane [6, 101, 149], or there may be large protein molecules in the membrane which 'rotate' across the membrane [27]. References to some of the transport models which have been proposed are given by Curran & Schultz [42], and it is not necessary here to explore the numerous possibilities in detail. In view of its descriptive simplicity the moving carrier model has much to recommend it, and it has been widely adopted in the interpretation of experimental observations.

CHAPTER 3

Kinetic theory of carrier transport

Experimental values obtained from carrier transport experiments have been related mathematically. The equations derive from a model in which certain fundamental assumptions are made [154, 159] and have been found in general to give a useful framework for the study of many aspects of transfer.

A theory of the kinetics of carrier transport will now be developed in stages, first by considering the attachment of the solute, or 'substrate' to the carrier in terms of adsorption, and then its transfer across the membrane. ('Substrate' is used here to mean a solute whose molecules are able to combine reversibly with adsorption sites.)

The relationship between an adsorption surface (which can be looked upon kinetically as a separate 'compartment') and adsorbed molecules can be considered in terms of physical chemistry. Molecules may become adsorbed on surfaces as a result of a residual field of force at the surface [49], and heat, known as 'heat of adsorption' [57] will be released in the process. If the system is thermodynamically isolated some of the heat will be transmitted to the molecules of the surface and will increase their mean free energy, resulting in increased agitation of molecules adsorbed. In some cases local energy will be sufficient to enable some adsorbed molecules to become detached, losing heat in the process.

Langmuir suggested a relationship between a gas and an adsorption surface at equilibrium which was strikingly supported by experiment [74], and which is expressed in the abbreviated form below:

$$\theta = \frac{\alpha\mu}{\alpha\mu + \nu} \tag{15}$$

in which θ is the fraction of the surface which is covered with gas molecules at equilibrium, μ is the number of gram-molecules of gas striking each unit area of surface per unit time, α is the (constant) proportion which adheres. $\alpha\mu$ then represents the rate at which gas molecules condense on the surface. ν is the rate at which the gas would evaporate if the surface were completely covered. We shall develop this equation from Langmuir's argument, but using different symbols by replacing μ by [S] and ν by $b.Q_{max}$. The meaning of these symbols will be explained below.

Let us assume that a proportion of the molecules of a substrate in a dilute solution which are continually striking an adsorptive surface as a result of thermal agitation will, in unit time, become attached to sites or regions on the surface. Of those which become attached a certain proportion may be expected to dissociate or leave the sites in unit time as a result of the agitation imparted by thermal energy. The number dissociating will be a function of the stability of the complex; it will thus be a proportion of the number attached to sites and will not be related directly to the number in solution. An equilibrium will be achieved in which the number of molecules being adsorbed in unit time will be equal to the number dissociating, and the number attached at any instant of time becomes a constant.

Assuming that an equilibrium of this sort has been reached, it is possible to relate the total number of molecules present, the proportion of surface occupied, and the number of molecules which are attached and detached in unit time. Suppose that the symbol m represents the number (or mass) of molecules in the solution which strike an unoccupied adsorption surface in unit time as a result of random movement. (Its value depends partly on the concentration of molecules in solution and partly on the area of the

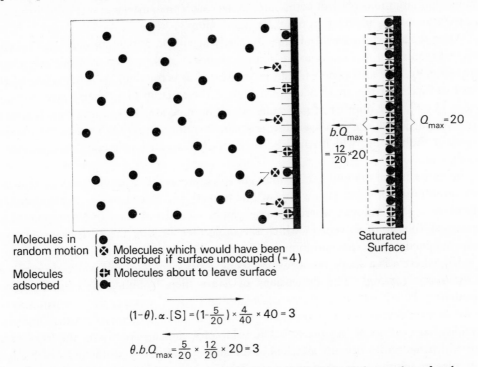

Molecules in random motion | ● Molecules adsorbed | ●

⊗ Molecules which would have been adsorbed if surface unoccupied (= 4)

⬌ Molecules about to leave surface

Saturated Surface

$Q_{max} = 20$

$b.Q_{max} = \frac{12}{20} \times 20$

$$(1-\theta).\alpha.[S] = (1-\tfrac{5}{20}) \times \tfrac{4}{40} \times 40 = 3$$

$$\theta.b.Q_{max} = \tfrac{5}{20} \times \tfrac{12}{20} \times 20 = 3$$

Fig. 9. The kinetics of adsorption at equilibrium. **Attachment.** *Left*: number of molecules in unit volume $= 40$ ($= [S]$). Number of molecules striking adsorption area (*broad vertical line*) and which would become attached in unit time if surface were unoccupied $= 4$, which can be expressed as the fraction $4/40$ ($= \alpha$) of the number in unit volume. Sites already occupied $= 5$, which can be expressed as the fraction $5/20$ ($= \theta$) of the total number of sites available (20). Sites vacant $= 15$, which is the fraction $15/20$, or $(1 - (5/20))$, ($= 1 - \theta$) of the total available. Number of molecules actually being attached in unit time is this fraction of the maximum possible, i.e. $15/20 \times (4/40 \times 40) = 3$, or $(1 - \theta).\alpha.[S]$. **Detachment.** *Right*: Maximum number of molecules that could be attached to adsorption surface $= 20$ ($= Q_{max}$). If surface were fully occupied, 12 molecules would dissociate in unit time ($= b.Q_{max}$), which is the fraction $12/20$ ($= b$) of that maximum. *Left*: number of molecules actually attached to adsorption surface $= 5$, which is the fraction $5/20$ ($= \theta$) of the maximum possible. Number actually dissociating in unit time is therefore $5/20 \times (12/20 \times 20) = 3$, i.e. $\theta.b.Q_{max}$. Thus at equilibrium, when the number becoming attached in unit time equals the number dissociating in unit time,

$$\frac{15}{20} \times \left(\frac{4}{40} \times 40\right) = \frac{5}{20} \times \left(\frac{12}{20} \times 20\right), \text{ or } (1-\theta).\alpha.[S] = \theta.b.Q_{max}.$$

(*Black circles* represent individual molecules; some are marked according to their behaviour, although there is nothing special about these.)

surface available.) A proportion, r, of these will become attached in unit time. The actual number or mass becoming attached is then $r.m$. The fraction, r, of the total is a measure of the 'readiness' with which molecules adhere to the surface, and depends on (i) the 'attraction' between the molecules and the adsorbing surface, and (ii) physical forces, such as temperature, which affect the rate of movement of molecules and hence the chance that any molecule will come into contact with the surface in a specified time. It has the dimensions of 'fraction/time', or, more simply, 'time^{-1}'.

When the adsorbing surface is already partly occupied, fewer molecules will become attached than when it is unoccupied, since occupied regions are not available for attachment (Fig. 9). The area occupied can be expressed as a fraction, θ, of the whole of the active surface, and the area unoccupied is then the fraction of surface which remains, that is, $(1-\theta)$. The number of molecules which become attached in unit time will be in direct proportion to the area available, and can therefore be represented as the fraction $(1-\theta)$, of the number which could become attached to an unoccupied surface, that is $(1-\theta).r.m$.

For convenience the number (or mass) of substrate in the expression can be replaced by 'concentration of substrate', which will be termed [S]. Since mass = concentration × volume, $m = [S] \times$ volume, where 'volume' can be considered as a constant representing the virtual volume of solvent in which the molecules striking the surface would have been contained at the concentration ruling; $r.m$ can then be written $(r.[S] \times$ volume$)$, or $\alpha.[S]$, where α is a single rate constant replacing $(r \times$ volume$)$ and which we may call an *association constant*. The dimensions of α are then 'volume/time'. Since we are considering the adsorption of solute molecules and not of solvent, the loss can also be shown as a reduction in concentration in the solvent, and is therefore analogous to the physiological concept of *'clearance'*.* The value of α depends partly on the 'readiness' with which molecules become attached (see r above) and partly on factors which allow molecules in solution at a particular concentration to reach (and strike) the surface at a critical rate. These factors are: (i) viscosity of fluid and hence limitation of movement of molecules, and (ii) any device, such as agitation of the solvent, which minimizes the concentration gradients which might otherwise exist in the solution.

It is also possible to express the rate at which molecules already attached to the adsorbing surface become detached. If the mass or quantity of molecules adsorbed at any instant of time is Q, the maximum which the surface can hold is Q_{max}, when the surface is said to be 'saturated'. (Q and Q_{max} can be expressed as the mass of substrate under observation which is attached per unit area of adsorbing surface, and it is assumed

* *Clearance* is the volume of solvent (at the observed concentration) whose concentration of solute would be reduced to zero by the observed loss. That is to say, if m is the mass of solute lost in unit time, and [S] is the concentration of the solute from which this material disappears, the volume 'cleared' will be $m/[S]$ units of volume per unit time. The reader may argue that there must be some criterion by which we choose the value of [S]; is it before m is lost, after m is lost, or at some point in between? Frequently, in many fields of experimental work, this loss is measured in the steady state; that is to say, a quantity of solute equal to m is continuously administered so that the value of [S] does not change. In those instances in which clearance is measured in the non-steady state it is usual to take the geometric mean of the values of [S] at the beginning and end of the particular unit of time considered.

throughout that adsorption occurs as a single layer of molecules. However, with certain adsorption systems, and in many systems in some conditions, molecules may form more than one layer on the surface as saturation approaches, so that more molecules are adsorbed than might have been expected [57].) In unit time a fraction, b, of the molecules adsorbed become detached. This fraction can be regarded as a *dissociation constant*, and depends on the stability of the complex; it has the dimension of 'fraction/time', or 'time^{-1}'. The number of molecules which leave a *saturated* surface in unit time is then $b.Q_{max}$, which is a rate constant for any specified substrate-surface combination, and has the dimensions of 'mass/time'. When a fraction, θ, of the surface is occupied, the number or mass of molecules then leaving in unit time is that fraction of the mass which would leave a saturated surface, that is, $\theta.b.Q_{max}$.

At equilibrium the number of molecules becoming attached in unit time is equal to the number leaving, by definition, so that:

$$(1-\theta).\alpha.[S] = \theta.b.Q_{max}, \tag{16}$$

the dimensions of both sides of the equation being 'mass/time'.

This equation can be rearranged so as to define the proportion of adsorption surface which is occupied, that is, θ, or the 'proportionate saturation', thus:

$$\theta = \frac{\alpha.[S]}{\alpha.[S]+b.Q_{max}}, \tag{17}$$

which can be rearranged to:

$$\theta = \frac{[S]}{[S]+\dfrac{b.Q_{max}}{\alpha}}. \tag{18}$$

For a specified system the expressions $b.Q_{max}$ and α are constants: Q_{max} is a constant property of the adsorption surface in terms of the mass of a specified substrate which it can hold, and α and b are association and dissociation constants respectively.

The expression $b.Q_{max}/\alpha$ is therefore a constant, and corresponds to the 'equilibrium constant', or K_s, of physical chemistry [57], so that the equation can now be written:

$$\theta = \frac{[S]}{[S]+K_s} \tag{19}$$

From equation (18) the dimensions of K_s are

$$\frac{mass/time}{volume/time}, \text{ or } \frac{mass}{volume},$$

that is, 'concentration'.

The equation itself is of the general form $y = x/(x+a)$ which describes a rectangular hyperbola, and plotting θ against [S] will accordingly produce a curve such as that shown in Fig. 10. As the concentration of substrate increases, the proportion of adsorption sites occupied increases asymptotically towards saturation, so that when the concentration is very high indeed, θ approaches 1·0.

c

At low concentrations relatively few sites are occupied and an increase in the concentration increases the number of sites occupied almost proportionately. However, as more of the surface is covered with substrate molecules, there is increasingly less space available for the attachment of further molecules and the number dissociating increases. Thus at high concentrations a small change in concentration produces little difference in the proportionate saturation.

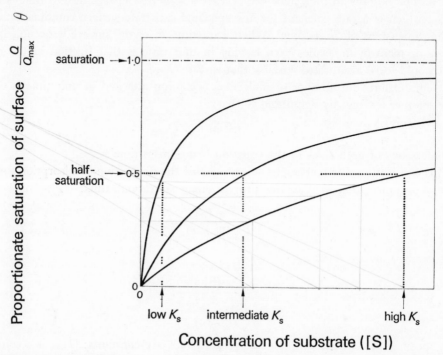

Fig. 10. Progressive saturation of adsorption surface with increase in concentration of substrate. K_s, the equilibrium constant, has a value equal to the concentration at which the surface is half-saturated. When the value of K_s is lower, a greater proportion of available sites is occupied at all concentrations of substrate than when it is higher. By adjustment of the scale of the abscissa all curves can be made to appear the same, since proportionate relationships are the same for all curves (eqn. 19).

It may be useful to replace θ by symbols more descriptive of the amount of substrate on the adsorption surface. Since Q and Q_{max} represent the mass of substrate adsorbed on an unsaturated and a saturated surface respectively, Q/Q_{max} represents the proportion of surface occupied, and hence is identical with θ. The equation can thus be written:

$$\frac{Q}{Q_{max}} = \frac{[S]}{[S]+K_s}. \tag{20}$$

Since K_s represents the relationship between association and dissociation, it determines the proportion of surface occupied at different concentrations of substrate. If the concentration of substrate is made equal to the value of K_s, the equation becomes:

$$\theta\left(= \frac{Q}{Q_{\max}}\right) = \frac{K_s}{K_s+K_s} = \tfrac{1}{2}. \tag{21}$$

The surface is then half-saturated, and the demonstration of the concentration which results in half-saturation would give the value of the parameter K_s.

The relationship can also be shown by considering first a surface which is half-saturated, so that θ and $(1-\theta)$ are both equal to $\tfrac{1}{2}$. At equilibrium the rates of association and dissociation are equal, so that

$$\tfrac{1}{2}.\alpha.[S] = \tfrac{1}{2}.b.Q_{\max} \tag{22}$$

i.e.

$$\alpha.[S] = b.Q_{\max} \tag{23}$$

hence

$$[S] = \frac{b.Q_{\max}}{\alpha} = K_s. \tag{24}$$

Thus, whatever the value of K_s, the proportionate saturation under these conditions (i.e. when $[S] = K_s$) is the same, that is, $\tfrac{1}{2}$. Consequently a substrate A whose K_s has a lower value than that of another substrate, B, will at any given concentration achieve a greater proportionate saturation than will substrate B, or will achieve the same proportionate saturation at lower concentrations, as shown in Fig. 10.

The value of K_s gives no information about the values of its component parts i.e. $Q_{\max}$, α, and b. For any given adsorption surface $Q_{\max}$ is a constant which is the same for all substrates, if we assume a mole for mole substrate-site relationship, for it then reflects only the number of sites available. The values of the association and dissociation constants, α and b, depend on the relationship between substrate and carrier. The value of K_s is influenced by both these constants, so that a difference in its value for different substrates may be due to a difference in the rate of association (or 'affinity'), a difference in the rate of dissociation, or a difference in both.

Carrier transport

A carrier which transports a substrate across a biological membrane may be looked upon for the purpose of kinetic description as a number of mobile adsorption sites, although various other models have been suggested [28, 42, 101, 148, 149]. Substrate is adsorbed at one side, moved across the membrane, and released at the other.

In the simplest case [153] the number of carrier sites at each surface of the membrane is assumed to be equal, and on each side the sites are effectively in adsorption equilibrium with substrate in solution on that side before moving across the membrane. It is assumed that equal numbers of sites move in each direction in unit time, that substrate molecules adsorb on to sites in a mole for mole relationship, and that the presence of attached substrate does not change the rate of movement of sites.

In a more complex but generalized hypothesis it has been suggested that the energy requirements for the different parts of the process may differ so that 'loaded' carrier sites may move across the membrane at a different rate from 'unloaded' sites [153]. This would result in an unequal distribution of sites. For example, 'loaded' sites may move faster than 'unloaded' ones, so that they would then move faster away from a surface where the concentration of substrate in solution was higher, than from a surface where it was lower or zero. It would follow that in any unit of time there would be more carrier sites present at the low-concentration surface than at the high-concentration surface. In a situation such as this transport would differ from one in which sites were equally distributed. There is some evidence that 'loaded' carriers may move faster than 'unloaded' ones [85, 86], although relevant experimental data may perhaps be explained in other ways (Chapter 7).

In evolving a kinetic theory which is to be tested it seems sensible to employ the simplest kinetic model as long as it is subsequently supported by the experimental data which have accumulated, and so it is with the simpler model that we shall mainly be concerned.

In living tissue, substrates are present on both sides of a membrane, but for simplicity we may assume that a substrate is initially in solution on one side only, and that over the period of observation the concentration on the other side is effectively zero. The nearest approach to this in practice would be a cell containing substrate but surrounded by substrate-free fluid of large volume. We also assume that equal numbers of carrier sites move across a membrane in each direction at equal rates, from which it follows that the total number of sites at each surface will not change, and it is to this constant number of sites that substrate is presented.

We look upon carrier sites as mobile adsorption sites which stay at each membrane surface long enough to come into virtual equilibrium with substrate before moving to the opposite surface [153]. The proportionate saturation of sites at the surface of the membrane can then be expressed in the same way as for adsorption equilibrium although, since substrate molecules are being continually removed, we actually have a steady state. The equation describing this is similar to that for adsorption, except that $b.Q_{max}$ is replaced by $\beta.Q_{max}$, the rate of dissociation of molecules (β) from a saturated carrier system while sites are in virtual equilibrium at the membrane surface. K_s or $b.Q_{max}/\alpha$ is replaced by K_m or $\beta.Q_{max}/\alpha$, although the two expressions can be virtually synonymous.

The proportionate saturation of sites achieved at one surface may be significantly affected by transfer and the dissociation at the other. This dissociation must then be incorporated as a component of K_m, effective dissociation at the original surface being represented as $(\beta.Q_{max}+\gamma.Q_{max})$ where γ is the fraction of substrate molecules which dissociate in unit time at the far surface of the membrane. K_m would then appear as $(\beta+\gamma)Q_{max}/\alpha$. This is a general expression of which $\beta.Q_{max}/\alpha$ is a special case in which the value of γ is negligibly small compared with the value of β. However, it is usually assumed in transport kinetics that carrier sites are in virtual equilibrium with substrate before leaving a membrane surface [153]; evidence suggests that steady state may be achieved very rapidly [24, 65], which would support this view.

The relationships just described are similar to those in enzyme kinetics, that is, the expression $\beta.Q_{max}/\alpha$ corresponds to the K_m of Michaelis theory [50, 97], and the expression $(\beta+\gamma)Q_{max}/\alpha$ corresponds to the K_m of Briggs–Haldane theory [23, 50], while α/Q_{max} corresponds to the enzyme rate constant k_1, β to k_2 and γ to k_3 [50]. (Although the term K_m has been widely adopted for use in transport kinetics, other terms, such as 'half-saturation constant' [141] and K_t [42, 165] have also been suggested but are not so commonly used.)

The equation describing the proportionate saturation of carrier sites in steady state with substrate at one surface of a membrane, and which is comparable with that describing adsorption (eqn. 19), is, then:

$$\theta = \frac{[S]}{[S]+K_m}. \tag{25}$$

Transported molecules of substrate which have dissociated at the far side of the membrane are clearly those which can be said to have been 'transported'; the mass transported in unit time is referred to as the transport rate, or v. The value of v depends on the total number of carrier sites, the number of molecules which can be adsorbed on to each site, and the rate of turnover of sites. If all sites are occupied before leaving the membrane surface there will then be a maximum quantity or mass of substrate molecules, that is Q_{max}, available for transfer. In unit time only a proportion of these will dissociate on the far side of the membrane, dissociation being assumed to occur exponentially within the unit of time of turnover. This proportion has already been referred to as γ.

The maximal rate of transport is thus $\gamma . Q_{max}$, and this can be abbreviated to v_{max}. It can now be said that the rate of transport is a fraction, θ (equivalent to the proportionate saturation), of the maximal rate, thus:

$$v = \theta . v_{max} \tag{26}$$

Substituting into this the expression for θ from equation (25) we have:

$$v = \frac{v_{max} . [S]}{[S]+K_m}. \tag{27}$$

This, the Michaelis–Menten equation [97], is the simplest of those equations which describe this model of transport by a carrier and clearly resembles equation (20) (p. 22) which describes adsorption. Transfer of this sort is usually referred to as 'following Michaelis–Menten kinetics', since it was first used by Michaelis and Menten [97] to describe enzyme reaction rates. It has also been used to describe those pharmacological reactions in which the same basic principles appear to apply [4]. It has, of course, the same algebraical and graphical properties as the adsorption equation.

Diffusion may contribute to transfer; the total rate of transfer is then the sum of the separate rates, thus:

$$v = \underset{\text{(carrier transport)}}{\frac{v_{max} . [S]}{[S]+K_m}} + \underset{\text{(diffusion)}}{K_D . [S]}. \tag{28}$$

The kinetic equation in relation to experimental work

The transfer of many substrates has been found to follow Michaelis–Menten kinetics in tissues such as the kidney (where the symbol 'T_m' corresponds to v_{max}) [142],

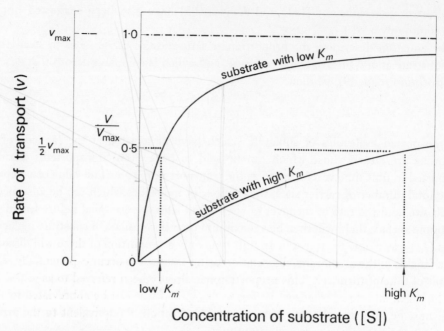

Fig. 11. Relationship between rate of transport and Michaelis constant (K_m). Transport may be expressed as a rate or as a proportion of v_{max} (i.e. v/v_{max}, which is equivalent to θ, or the proportionate saturation of sites). The Michaelis constant has a value equal to the concentration of substrate at which half the maximal rate of transport is developed, and hence at which the carrier system may be assumed to be half-saturated. A substrate with a higher value of K_m occupies less sites and hence is transported less rapidly at a particular concentration than one with a lower value of K_m. (Compare with Fig. 10.)

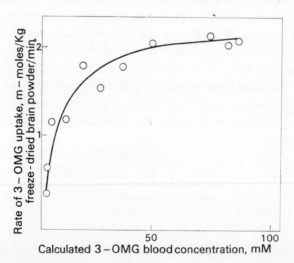

Fig. 12. Saturation of transfer with increase in concentration of substrate. Uptake *in vivo* of 3-O-methyl-D-glucopyranose (3-OMG) by the brain of the rat from various concentrations in the blood. Time, 5 min. (Redrawn from data published by Bidder[13].)

intestine [165], brain [104] (Fig. 12), erythrocyte [98, 99, 100, 154] as well as in a number of other tissues [29, 42, 69, 153, 159]. The relationships between transfer rate and concentration are shown in Fig. 11 and are similar to the general relationships shown in Fig. 10; an experimental example is shown in Fig. 12.

When the concentration of substrate is very low or very high *relative to K_m* certain simplifications can be made. For example, if the concentration is very low, the rate of transport is virtually proportional to the concentration of substrate, and hence it resembles a diffusion system. That is to say, if [S] is small enough *relative to K_m* the expression $([S] + K_m)$ approaches the value K_m and the kinetic equation (eqn. 27) approximates to:

$$v = \frac{v_{\max}}{K_m} \cdot [S] \qquad (29)$$

in which $v_{\max}/K_m$ is analogous to the diffusion constant K_D. At such concentrations the proportion of sites occupied is such a small proportion of the total number of sites that the probability of attachment of each molecule remains virtually the same in the presence of small changes in the concentration of substrate. Thus the number of molecules attached varies in proportion to the concentration. Conversely, if the concentration of substrate is very large *relative to K_m* the carrier is virtually saturated, and a small increase in concentration makes no detectable difference either to the number of molecules attached or to the rate of transport, which therefore reaches a virtually constant upper limit. That is to say, when $[S] \gg K_m$,

$$v \simeq \frac{[S]}{[S]} \cdot v_{\max} = v_{\max}. \qquad (30)$$

This difference in behaviour in different regions of a single kinetic curve, i.e. those of low and high concentration, has led to an artificial division of transfer kinetics into two ostensibly distinct types known as 'D-kinetics' on the one hand, representing diffusion-type kinetics, and 'E-kinetics' on the other, representing enzyme-type or saturation kinetics [135, 159]. This distinction may be misleading if it is not appreciated that both types in reality represent behaviour of a single kinetic type at low and high concentrations of substrate respectively *relative to K_m*.

When comparing quantitatively the transport of different substrates it may be of only limited value to carry out investigations over a narrow range of concentrations only. For example, it has often been said in commenting on experimental data under such conditions, that one substrate is transported 'better' or 'worse' (or some similar wording) than another substrate. Within the limits of a particular experiment this may be a reasonable statement, but it should not be assumed to apply beyond those limits without further evidence. Rate of transport is a reflection of the kinetic parameters of a particular substrate-carrier complex and relationships between two substrates may differ markedly at widely separated concentrations. This is illustrated in Fig. 13. The two curves represent the rates of entry of the two amino acids L-leucine and L-valine into erythrocytes, calculated from published kinetic constants [163]. It might be said from experiments in which each amino acid was present at a concentration of about 4 mM that they were both transported to the same extent. However, at concentrations less than this, L-leucine would be transported more rapidly than L-valine, while at concentrations above it, L-valine would be the more rapidly transported. These differences are a reflection of the fact that L-leucine has a lower $v_{\max}$ and a lower K_m than L-valine. It is clear that the relative transport rates here depend on the concentration at which the observations are made.

The relationships can be shown algebraically. Thus, if v_1 and v_2 are the rates of transport of Substrates 1 and 2 respectively, it follows from equation that:

$$\frac{v_1}{v_2} = \frac{\dfrac{v_{\max_1} \cdot [S]}{[S] + K_{m_1}}}{\dfrac{v_{\max_2} \cdot [S]}{[S] + K_{m_2}}}. \tag{31}$$

If $[S]$ is large in relation to K_{m_1} and K_{m_2} this approximates to:

$$\frac{v_1}{v_2} = \frac{v_{\max_1}}{v_{\max_2}} \tag{32}$$

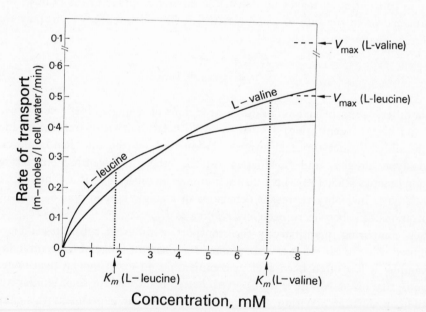

Fig. 13. Effect of concentration on transport relationship between two amino acids, as shown by the calculated rate of entry of L-leucine and L-valine into the human erythrocyte. At a concentration of about 4 mM the rate of transport of the two amino acids would be similar, but below this concentration L-leucine, and above it, L-valine would be the more rapidly transported. The curves have been drawn from values calculated by inserting the following published kinetic constants into equation (27). L-leucine: $v_{\max} = 0\cdot52$ mmole/l cell water/min, $K_m = 1\cdot8$ mM; L-valine: $v_{\max} = 1\cdot0$ mmole/l cell water/min, $K_m = 7\cdot0$ mM (incubation < 1 min at 37°C) [163].

and if [S] is small in relation to K_{m_1} and K_{m_2} it approximates to:

$$\frac{v_1}{v_2} = \frac{v_{\max_1}.K_{m_2}}{v_{\max_2}.K_{m_1}}. \tag{33}$$

It follows that a substrate which has a higher $v_{\max}$ and a higher K_m than another may, depending on the quantitative relationships of the constants, have a relatively higher rate of transport at high concentrations (eqn. 32) and a relatively lower rate of transport at low concentrations (eqn. 33), as in the example illustrated in Fig. 13.

$$K_m$$

The constituent parts of K_m have already been discussed, but from the experimental point of view in terms of known substrate concentrations and observed transport rates it will be seen that the value of K_m is equal to that concentration of substrate at which the transport rate is exactly half the maximal transport rate (cf. K_s for adsorption).

K_m has often been referred to as a constant describing 'affinity' [29, 159, 165], a low value for K_m being associated with a high 'affinity' of substrate for carrier and a high value with a low 'affinity'. 'Affinity' has been defined as a 'tendency of certain bodies to unite with others' [166] and implies only *association*. Certainly on these grounds K_m could be said to be inversely proportional to affinity provided that different substrates have the same *dissociation* constant, but to describe K_m only in terms of affinity is liable to lead to misinterpretation. As Riggs [128] has said, K_m is 'not a rate constant, nor an affinity constant, nor a dissociation constant, but merely a constant of convenience'.

An alternative way of regarding K_m is in relation to $v_{\max}$, when the ratio $v_{\max}/K_m$ is seen to be a clearance (p. 20) which, in turn, is just another 'constant of convenience'.

The ability of substrates to associate or dissociate may be expected to differ one from another, owing to differences in their physicochemical relationships with a carrier, and would be shown by differences in the value of K_m for each. Values may differ by as much as 300-fold when comparing different substrates and different tissues [77, 98, 157], but the absence of any difference in K_m does not necessarily indicate that there is no difference in the rate constants whose ratio is K_m.

For the transfer of sugars by the erythrocyte the value of K_m is closely related to the three-dimensional shape of the sugar molecules [80]. In other systems transfer and hence possibly K_m can be influenced not only by the shape of the substrate molecule, but also by the location and nature of its ionic charge [108]. There is also evidence that the presence of other charged ions can affect the value of the K_m of some substrates [40, 139a].

The separate components of K_m cannot easily be determined. Whereas in enzyme kinetics it is sometimes possible to determine the comparable components by the use of purified enzyme [50], carriers cannot be similarly investigated even if they do have a 'discrete' existence and even if they are intramembrane enzymes; it is difficult to see how they could be purified without the loss of transport identity which would result from separation from the membrane. If, on the other hand, they are integral parts of membrane structure, the experimental examination of a coordinated system in sub-total disarray might be expected to have only limited reward.

$$v_{max}$$

When a component of transfer is found to behave according to Michaelis–Menten kinetics, a value for v_{max} is often found by extrapolation of values obtained over the experimental range [9, 16, 68, 114].

Its value is a function of two factors: firstly, the capacity of the carrier system (Q_{max}) and, secondly, the proportion of adsorbed molecules which dissociate in the 'forwards' direction in unit time (γ). The value of Q_{max} depends on (i) the number of varrier sites per unit area of membrane, (ii) the number of substrate molecules which can be adsorbed on to each site, and (iii) the number of carrier-containing regions of membrane in a specified mass of tissue, or, where it can be determined, in a specified area of membrane. If there is a mole for mole substrate-carrier site relationship, all substrates transported can be expected to have the same value for Q_{max}. The value of γ depends on two factors: (i) the probability that a substrate molecule will dissociate from a carrier site in a given time, and (ii) the rate of turnover of carrier across the membrane. If the rate of turnover is relatively slow, the value of γ, and hence of v_{max}, can be expected to be primarily related to turnover, and hence to be similar for various substrates. If it is relatively rapid, the value of v_{max} would be more affected by variations in the factors affecting dissociation itself, and these may not be the same for each substrate. In enzyme kinetics, for example, v_{max} may vary from one substrate to another when applied to the same enzyme [50], but in view of differences in the nature of the organization of the two systems, too close a parallel must not at present be drawn.

It is commonly accepted that it is the movement of carrier which is the limiting factor for v_{max} [40, 77, 153, 154]. This may indeed be so, for example, with the transport of sugars across the erythrocyte membrane, since a number of them (arabinose, galactose, glucose, ribose and xylose) appear to have the same maximal rate of transfer [77, 98], although not all authors agree that this is so [157]. On the other hand, in brain, v_{max} varies for different amino acids, although certain groups of amino acid appear to show approximately the same value [16], and this has been attributed to the existence of a common carrier for each group [16, 107]. Interpretation is confused, however, by evidence that the transport of a number of amino acids is performed by more than one carrier system [107].

Net carrier transport

When a substrate is present on both sides of a membrane the simple kinetic analysis above must be modified, since transfer then takes place in opposite directions. This is not only a much more realistic position, but is the general model of which one-way transport is a special case. Although carrier transport is extensively studied as if it consisted only of entry of substrate into cells, it is clear that substrate is also moved out [31, 76, 81, 82].

When a substrate-free cell is placed in a medium containing substrate which can diffuse or be transported into the cell there is, for a brief time only, effectively one-way transfer. Substrate molecules which enter the cell may be distributed rapidly throughout the cellular compartment available to them. Some of these molecules will impinge upon the interior surface of the membrane, so that there will then (and subsequently) be an outward movement of molecules either by carrier or by diffusion or both.

The rate of carrier transport in each direction (v and v') is expressed thus:

$$v = \frac{v_{\max} \cdot [S]}{[S] + K_m} \qquad v' = \frac{v_{\max}' \cdot [S']}{[S'] + K_{m'}} \qquad (34, 35)$$

(inward transport)   (outward transport)

where v and v' are the inward and outward rates respectively, $[S]$ and $[S']$ the extra-cellular and intracellular concentrations of substrate, $v_{\max}$ and $v_{\max}'$ the inward and outward maximal rates, and K_m and $K_{m'}$ the Michaelis constants for inward and outward transport. (To avoid ambiguity of direction, we shall hereafter refer to 'inward' and 'outward' in this way.)

Net transfer, including diffusion, can then be expressed as:

$$V = (v - v') = \left(\frac{v_{\max} \cdot [S]}{[S] + K_m} - \frac{v_{\max}' \cdot [S']}{[S'] + K_{m'}} \right) + K_D([S] - [S']), \qquad (36)$$

where V represents the net rate of transfer; a positive sign indicates that net transfer is inwards, a negative sign that it is outwards.

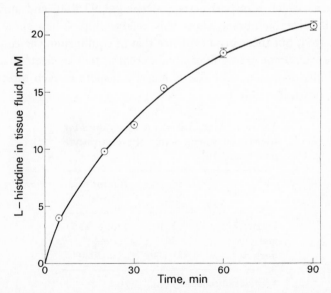

Fig. 14. Change in intracellular accumulation of substrate with time. Concentrations of L-histidine by mouse brain slices after various incubation times, shown as mean and standard error of mean (if not shown, standard error lies within limits of symbol); $n \geqslant 5, \leqslant 20$. Initial concentration of L-histidine in suspending medium, 2 mM, 37°C. (Drawn from previously published data [104].)

As the concentration of substrate inside the cell increases, the rate of movement outwards increases and continues to do so until eventually the intracellular concentration is high enough for the outward rate to equal the inward rate, so establishing an equilibrium. If the intracellular concentration is plotted against time, it will be seen to rise at an ever-decreasing rate (Fig. 14) until at equilibrium it rises no further. The relationship between concentration and time can be expressed mathematically along the same lines as was shown for diffusion (p. 8) but is more complex; here the constant, k, of equation (12) is replaced by a variable, $v_{max} . [S]/([S]+K_m)$, where $[S]$ is changing with time.

If at equilibrium the concentrations on each side of the membrane are the same (ignoring inequalities of concentration due to the effects of non-specific physical forces such as the Donnan effect [20, 45] or static charge [139a]) carrier transport is of the 'equalizing' type. If the concentrations are different it is of the 'concentrative' type.

Equalizing transfer

When the kinetic constants of inward transport are the same as those of outward transport, the transport rate is the same in each direction at any stated concentration in either compartment. At equilibrium net transport is zero ($V = 0$), and the equation (eqn. 36) above (omitting diffusion) becomes:

$$\frac{v_{max} . [S]}{[S]+K_m} = \frac{v_{max} . [S']}{[S']+K_m},\tag{37}$$

i.e. $[S] = [S']$.

Carrier systems which transport certain sugars [98, 99, 100] and amino acids [163] across the erythrocyte membrane show this relationship. They show properties of saturation [154, 163], but there is no evidence that at equilibrium the concentrations on either side of the membrane are significantly different or that an electrochemical gradient is established. Individual values for K_m of sugar transport in each direction appear to be the same, as shown in Table 1.

Table 1. A comparison of K_m values for transfer of sugars across the erythrocyte membrane [76]

Sugar	K_m for entry (mM)	K_m for exit (mM)
Dextrose	7·5, 10	7·5, 8, 8, 9
Galactose	50	44
Sorbose*	1300–2000	*ca.* 2000

* Values unreliable [76].

Concentrative transfer

We speak of concentrative transfer when the observed concentration of substrate on one side of a membrane is greater than that on the other at a time when net transfer is zero, and when the difference cannot be accounted for by inequalities due to non-specific physical forces or to 'countertransport' (described in Chapter 7). It is now well established that concentrative transfer of this sort can occur in a number of tissues [69, 108, 121, 159, 165]; for example, amino acids, for which there is evidence of outward carrier transport [31, 81], as well as of inward transport [69, 108] may develop a concentration within a cell which is greater than that outside by twenty times or more [17]. Interference with the energy metabolism of a cell greatly reduces the effectiveness of concentrative transport [69] and may even eliminate it [106], while the presence of Na^+ may be an essential feature [139a].

It will be apparent that if we make $V = 0$, equation (36) becomes (omitting diffusion):

$$\frac{v_{max} \cdot [S]}{[S] + K_m} = \frac{v_{max}' \cdot [S']}{[S'] + K_m'}. \tag{38}$$

If $[S']$ is to have a value greater than that of $[S]$, either K_m' must be sufficiently large in relation to K_m, or v_{max}' sufficiently small in relation to v_{max}; although differences

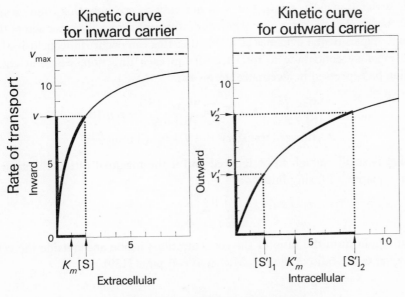

Concentration of substrate

Fig. 15. Attainment of a concentration gradient between the inside and outside of a cell. *Left: transport inwards* is by a carrier with a low K_m. *Right: transport outwards* is by a carrier with a higher K_m. Units are arbitrary. The rate of inward transport (v) is constant at a value of 8, assuming the extracellular concentration of substrate ($[S]$) to be constant at a value of 2. (*left*). As the intracellular concentration of substrate rises, the outward rate of transport rises (*right*). When the intracellular concentration has reached the same value as that outside ($[S']_1 = [S]$) the inward rate (v) is still greater than the outward rate (v_1'). Not until an intracellular concentration of 8 ($[S']_2$) has been reached does the outward rate (v_2') equal the inward rate (v); the intracellular concentration then stops rising Equilibrium has now been achieved with a higher concentration of substrate inside the cell than outside ($[S']_2 > S$).

may be described in this way, they infer a change in the carrier itself on each 'side' of the membrane.

$K_{m'}$ may indeed have a higher value than K_m, as has been found experimentally [27], and this may presumably be due to some change in the characteristics of the carrier sites at the inner surface of the membrane, comparable to that associated with the Bohr effect of haemoglobin [118]. The substrate would then become attached less readily at the inner surface of the membrane than at the outer surface. The manner in which concentrative transport may occur by this means is illustrated in Fig. 15. The left-hand diagram shows a kinetic curve for inward transport ($K_m = 1$, $v_{max} = 12$; arbitrary units), and the right-hand diagram shows a curve for outward transport ($K_{m'} = 4$, $v_{max'} = 12$, in arbitrary units). The extracellular concentration of substrate is fixed at a value of 2, so that the inward rate (v) is 8. When the concentration inside the cell has risen to that outside ($[S] = [S']_1 = 2$) the outward rate of transport is still less than the inward rate ($v'_1 = 4$), so that net transport is still inwards. Equilibrium (when $v = v'_2 = 8$) is not reached until the intracellular concentration is four times that outside ($[S']_2 = 8$).

The effective capacity of inward carrier transport, v_{max}, may be greater than that for outward carrier transport, $v_{max'}$, but it is not easy to see how this could come about except by saying that the rates of movement of different forms of carrier across the membrane vary. Experimental evidence as to whether this can occur is inconclusive [31].

Diffusion may contribute to total transfer in each direction, which at equilibrium would then be expressed in algebraical terms as:

$$\frac{v_{max} \cdot [S]}{[S] + K_m} + K_D \cdot [S] = \frac{v_{max'} \cdot [S']}{[S'] + K_{m'}} + K_D \cdot [S'] \tag{39}$$

$$\text{(inward transfer)} \qquad \text{(outward transfer)}$$

If diffusion is small enough to be ignored and if the maximal rate is the same in each direction, equation (39) simplifies to:

$$\frac{[S']}{[S]} = \frac{K_{m'}}{K_m}, \tag{40}$$

that is, at equilibrium the ratio of the concentrations inside and outside the cell would here be equal to the ratio of the two Michaelis constants [159].

Effect of two-way carrier transport on values of kinetic constants

In certain cases the value of K_m when measured during two-way transport across a membrane has been found to be higher than during one-way transport [86]. For example, K_m for one-way glucose transport across the erythrocyte membrane was found to be two or three times higher when glucose was present at equal concentrations on both sides of the membrane (and hence moving at equal rates in both directions) than when it was present on only one side.

To explain this let us assume that the carrier comes virtually to equilibrium with substrate on each side of the membrane before moving across to the other side. Remembering that in such a state the rate of association of substrate is equal to the rate of dissociation, the kinetic relationship can be expressed in the simple form similar to that for equilibrium (eqn. 16) thus:

$$(1-\theta).\alpha.[S] = \theta.\beta.Q_{max} \tag{41}$$

or

$$\theta = \frac{[S]}{[S]+\dfrac{\beta.Q_{max}}{\alpha}} = \frac{[S]}{[S]+K_{m_1}} \tag{42, 43}$$

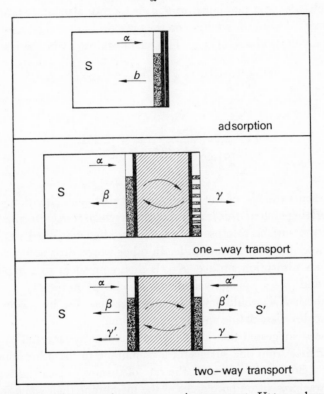

Fig. 16. Adsorption and one- and two-way carrier transport. *Upper:* adsorption surface in equilibrium with substrate. *Middle:* carrier transport from a compartment of limited volume into a compartment of large or unlimited volume. In compartment on *left*, substrate and carrier sites may be in virtual equilibrium before substrate is transported to *right*, where substrate dissociates exponentially as long as occupied sites remain there, a fraction, γ, dissociating in each unit of time. The larger the value of γ relative to that of β, the more will γ affect the proportionate saturation of carrier sites. *Lower:* equalizing carrier transport between two compartments of limited volume at equilibrium. In each compartment a proportion of the dissociating molecules has been derived from the opposite compartment, so that total dissociation at each surface involves a mixture of molecules from both compartments. *Broad vertical lines* represent adsorption surface comprising randomly distributed sites, fixed in the case of adsorption, mobile in the case of carriers. *Adjacent shaded areas* represent proportion of sites randomly occupied. *Cross-hatched areas* represent membrane, and *contained arrows* the movement of carrier across. α represents the association constant, b, β and γ dissociation constants, and S and S′ the substrate present.

where θ is the proportional saturation of carrier sites at each surface. These are the relationships described earlier for one-way transport (eqn. 25), when carrier sites return to substrate-bathed surface after being in effective equilibrium on the opposite side where substrate concentration is virtually zero. It is shown in the middle diagram of Fig. 16, and may be compared with the upper diagram which shows adsorption.

When there is substrate on both sides of a membrane the relationships are altered, as shown in the lower diagram of Fig. 16. In approaching equilibrium with the carrier sites, substrate molecules being adsorbed may occupy not only sites which are vacated by substrate molecules at a rate equal to $\theta.\beta.Q_{max}$, but also sites vacated by substrate molecules transported from the solution on the opposite side of the membrane and which dissociate at a rate equal to $\theta.\gamma'.Q_{max}$. The relationships can be expressed as:

$$(1-\theta).\alpha.[S] = \theta.\beta.Q_{max}+\theta.\gamma'.Q_{max}, \qquad (44)$$

which can be rearranged to:

$$\theta = \frac{[S]}{[S]+\dfrac{(\beta+\gamma')Q_{max}}{\alpha}} = \frac{[S]}{[S]+K_{m_2}}, \qquad (45, 46)$$

where K_{m_2} represents the K_m of the system in this particular situation.

In an equilibrating system, this is as if those sites which transport substrate molecules, i.e. which leave the surface to cross the membrane, had remained stationary and discharged those molecules on the same side. In other words, it is as if the carrier system had reverted to an adsorption surface. K_{m_2} is then equivalent to a hypothetical K_s for the system [86] and $(\beta+\gamma')$ is equivalent to b of adsorption (p. 21).

In this hypothesis K_{m_2} would have a value higher than K_{m_1} by reason of an increased number of molecules dissociating in any unit of time.

It has also been suggested that kinetic constants may be affected by the movement of loaded carrier sites from one surface of a membrane to the other more rapidly than unloaded sites [85, 86]. If this were so, the rate of turnover of carrier would be greater when substrate was present on both sides of the membrane than when it was effectively present on only one side, since more sites would be occupied in any unit of time and hence the mean rate of movement would be greater. It would result in an increase of v_{max} as well as of K_{m_1}, as has been found experimentally [85], but the findings could also be explained in terms of competition along the general lines described in Chapter 7.

It is also relevant at this point to mention some investigations in which a particular protein in the membrane of certain bacteria was believed to be associated with the transport of amino acids [119a]. By exposing the bacteria to osmotic shock by immersion in water, a protein appeared in the suspending fluid on to which the amino acids could be adsorbed. It was found that the value of K_s for adsorption was similar to that of K_m for transport, and this was claimed as evidence that the protein was associated with transport. However, on the hypothesis described above, K_s for adsorption might be expected to be higher than that for transport by the same sites [86], and this would suggest that the isolated protein was not solely responsible for transport.

The direction of diffusion in the
presence of carrier transport

The overall or net direction in which molecules move by diffusion is not necessarily the same as the overall direction of movement by carrier transport.

In equalizing transfer, diffusion is in the same direction as net carrier transport, for molecules migrate from a region of higher concentration to a region of lower concentration in each case. This is illustrated in Fig. 17; the rate of transfer by each component is shown together with total transfer.

In concentrative transfer, however, net diffusion may be in the opposite direction to net transport by the carrier, as shown in Fig. 18. When carrier transport is from a region of lower concentration to a region of higher concentration, net diffusion, which must be down the concentration gradient, will be in the opposite direction. In Fig. 18 it is assumed that at lower substrate concentrations the intracellular concentration is higher than the extracellular concentration so that diffusion is outwards (negative in value), but that at higher concentrations the intracellular concentration is lower and diffusion is then inwards (positive in value).

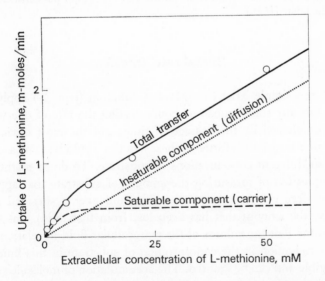

Fig. 17. Components of equalizing transfer, illustrated by the uptake of the amino acid methionine by erythrocytes at various extracellular concentrations. Total transfer has been divided by the original authors into saturable and unsaturable components which probably represent carrier transport and diffusion respectively, but they point out that it is still possible that the so-called unsaturable component may show evidence of saturation at concentrations above 100 mM (the highest concentration used, not shown here). Incubation not more than 5 min, 37°C. (Redrawn from data published by Winter & Christensen [163].)

D

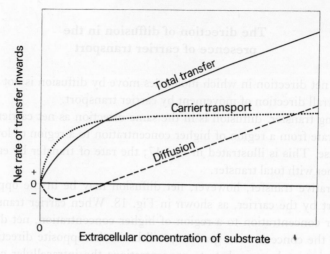

Fig. 18. Components of concentrative transfer shown as the sum of inward carrier transport and net diffusion; outward carrier transport is assumed to be negligible. At low extracellular concentrations of substrate it is assumed that the concentration in the cell becomes higher than that outside so that diffusion is outwards and therefore negative in value. At high extracellular concentration, diffusion is inwards and therefore positive in value. These relationships may apply under 'initial rate' conditions. In concentrative transfer, the intracellular concentration will always be higher than that outside at equilibrium so that diffusion will then be outwards. (This graph was constructed with the aid of published data [163].)

'Initial rate' transfer

The kinetics of the simple Michaelis–Menten equation (eqn. 27) apply to a one-way transport only. For any stated substrate concentration the rate of one-way transport is assumed to be unaffected by the passage of time, and the most obvious way to find whether a particular transfer conforms to these kinetics would be to measure the one-way rate of transfer at different concentrations of substrate. To do so might at first appear to be a simple question of measuring the amount of substrate that appears within an initially substrate-free cell from a surrounding fluid over a specified period of time (or, alternatively, the amount that has been lost from that fluid), but difficulties now arise. Let us assume that the volume of the extracellular fluid in relation to cell volume is so large that a reduction in the concentration of substrate in this fluid as a result of transfer is negligible and can be ignored. The accumulation of molecules which develops in the cell (usually used as a measure of 'transfer') represents not one-way transfer, but net transfer; except for the first moment of time it must incorporate also transport from within the cell outwards. Thus the ability to measure substrate within a cell is itself an indication that the assumption that we are measuring one-way transfer cannot literally be true.

Since movement in both directions will always exist however short the time interval used, the absolute rate of one-way transfer cannot be measured, and in practice a compromise has to be accepted. The net accumulation of substrate in a cell over a short time interval (say, one or two minutes) may be measured, and then treated as if it were a one-way rate. It is then called the 'initial rate', and its validity as a measure of one-way movement inwards is based on the assumption that the amount of substrate which has accumulated in the cell is small enough for outward transfer to be ignored.

Even with such a short time lapse as this there may be difficulty in two ways. In the first place inward transfer may be so rapid that there is substantial accumulation within the cell and hence outward transfer within the specified period may be considerable. In the second place, outward transfer may be so rapid that a somewhat longer time interval is required before reliable observations of intracellular concentration can be made. In the first circumstance the difficulty may be minimized by shortening the time interval, but in the second case little can be done to obtain reliable values by this method.

An example of the effect that rapid outward transport can have on net accumulation is shown in Fig. 19. This shows the net rate of uptake of the stereoisomers of the amino acid alanine over various time intervals. Initially L-alanine was accumulated to a greater

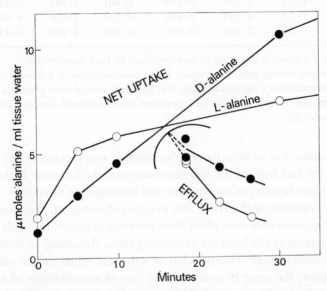

Fig. 19. Effect of rate of efflux (outward movement) on net accumulation by tissue. Labelled D-isomer or labelled L-isomer of the amino acid alanine was taken up by brain tissue from a suspending medium containing the appropriate isomer at 2 mM concentration. At times less than 16 min, the tissue had accumulated more L-isomer than D-isomer, but at times greater than 16 min it had accumulated more D-isomer than L-isomer (*net uptake*). When tissue was taken from the incubation medium at 16 min (i.e. when the tissue concentration of each isomer was the same) and was then incubated in medium containing the corresponding 2 mM unlabelled isomer, the labelled L-isomer moved out of the tissue more rapidly than the labelled D-isomer (*efflux*). Thus the L-isomer probably entered more rapidly than the D-isomer; at short incubation times, with the intracellular concentration low, net accumulation would reflect primarily influx, whereas at long incubation times, with the intracellular concentration higher, efflux contributed more to net accumulation. *Ordinate:* tissue concentration. *Abscissa:* incubation time. Incubation 37°C. (Redrawn from data published by Neame & Smith [110].)

Table 2. Calculated error in 'initial rate' measurements. Inward and outward rates of carrier transport calculated from published values of constants for transport of N-methyl-α-aminoisobutyrate by Ehrlich ascites carcinoma cells [31], using equation (36) but omitting diffusion component. Inward $K_m = 0 \cdot 3$ mM, outward $K_m = 5$ mM, v_{max} in each direction $= 1 \cdot 5$ mmoles/kg cell water/min.* External concentration = inward K_m. Error represents comparison of rate of accumulation with inward rate.

			Mean for previous minute					
Time (min)	External concentration (mM)	Inward rate (mM/min)	Assumed internal concentration† (mM)	Outward rate (mM/min)	Net inward rate (mM/min)	Cell concentration‡ (mM)	Rate of accumulation (mM/min)	Error (per cent)
	[S]	v	[S′]	v'	$v-v'$			
1	0·3	0·75	0·375	0·105	0·645	0·645	0·645	14§
2	0·3	0·75	1·020	0·254	0·496	1·141	0·571	24
3	0·3	0·75	1·516	0·349	0·401	1·542	0·514	31
4	0·3	0·75	1·917	0·416	0·334	1·876	0·469	37
5	0·3	0·75	2·251	0·466	0·284	2·160	0·432	42

* For calculations, 1 g tissue is assumed to be equivalent to 1 ml tissue water.
† Assumed mean value during previous minute (prior accumulation $+ \frac{1}{2}$ inward rate).
‡ Previous cell concentration + net inward rate (i.e. amount entered over previous minute).
§ Error varied between 13 per cent and 15 per cent when external concentration increased or decreased a hundred-fold.

extent than D-alanine, but as time passed the position was reversed. The difference can be attributed to the fact that, of the two stereoisomers, the L-isomer is the one which is moved outward more rapidly, as shown from the investigation of efflux.

A quantitative estimate of the error, due to outward transport, that may be associated with 'initial rate' experiments over short time intervals is shown in Table 2, using published kinetic constants as the basis for calculating rates. Assuming an initially substrate-free cell and an extracellular concentration of substrate equal to the value of the K_m for inward transport, the error in assuming the rate of accumulation of substrate to be equal to the inward rate of transport is 14 per cent for a one-minute period, while for a five-minute period it is 41 per cent. Although accumulation is not linear in relation to time [104, 114, 156], it is reasonable to assume that it is virtually so over very short time intervals, but for longer time intervals accumulation should not be expressed in terms of a uniform rate.

Chapter 4

Analysis of Experimental Data

Experimental data are often analysed with a view to determining whether they conform to simple Michaelis–Menten kinetics and, if so, what values may be found for the kinetic constants v_{max} and K_m. This chapter shows graphical and arithmetical ways in which this may be done. Some of the graphical methods are widely used because of their simplicity, although often at the expense of accuracy.

Graphical methods

If the contribution of diffusion to total transfer is assumed to be negligible or known, the basic equation (eqn. 27) for the carrier contribution is

$$v = \frac{v_{max} \cdot [S]}{[S] + K_m},$$

an equation which describes a rectangular hyperbola. Simple inspection of a data plot may often be inadequate, and to overcome this difficulty a rearrangement of the equation to give a straight line is commonly employed. If the experimental results do not then conform satisfactorily to a straight line, the criteria of the model have not been met and transfer cannot be accounted for solely in terms of the model.

There are several algebraical rearrangements of the above equation which produce a linear function. Unfortunately, it is deceptively easy to draw a straight line through almost any series of values on a graph, particularly if they are widely scattered or few in number. The use of an algebraical rearrangement of the data is also liable to distort or reduce awareness of experimental error [51, 128], and we suggest that those who would use these methods should consult appropriate works of reference.

We shall now discuss three methods of linear transformation in common use, all originally devised for use in the analysis of enzyme kinetics [50, 62].

The double reciprocal or Lineweaver–Burk plot [50, 62, 87]

This is the easiest of the three graphical methods to interpret (and hence is the most widely used) since it directly relates the reciprocal of the observed transfer rate to the reciprocal of the known substrate concentration.

The general form of the kinetic equation (eqn. 27) above is

$$y = \frac{mx}{x+c},$$

whose reciprocal is

$$\frac{1}{y} = \frac{x+c}{mx}$$

which in turn can be rewritten

$$\frac{1}{y} = \frac{c}{m} \cdot \frac{1}{x} + \frac{1}{m},$$

where c and m are constants. This is a linear equation, whose general form is $y' = ax' + b$. In a similar way, inversion of the kinetic equation (eqn. 27) clearly gives:

$$\frac{1}{v} = \frac{K_m}{v_{max}} \cdot \frac{1}{[S]} + \frac{1}{v_{max}}. \tag{47}$$

A plot of $1/v$ against $1/[S]$ is, therefore, linear. K_m/v_{max} is a constant representing the slope of the line and $1/v_{max}$ is the distance above the origin at which the line intersects the ordinate. An example of this is shown in Fig. 20. (The experimental points extend beyond the confines of the figure.)

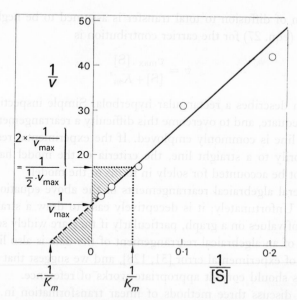

Fig. 20. Double reciprocal plot of the uptake of γ-aminobutyrate by brain slices. Shaded areas are similar triangles, showing that the plotted straight line cuts the abscissa at a value equal to $-(1/K_m)$. v = rate of uptake of γ-aminobutyrate, μmoles/min/g tissue, [S] = concentration of γ-aminobutyrate in suspending medium, mM. $v_{max} = 0.115$ μmoles/min/g tissue, $K_m = 22$ μM. Incubation 10 min, 25°C. (Adapted from part of data published by Iversen & Neal [64].)

K_m is obtained by recognizing that it is equal to the concentration of substrate at which the rate of transport is half-maximal, i.e. $\frac{1}{2}.v_{max}$. The reciprocal of $\frac{1}{2}.v_{max}$ is $2\times(1/v_{max})$ and is therefore represented on the ordinate at a distance from the origin which is twice that of $1/v_{max}$. The value of $1/[S]$ at this point on the graph is then K_m.

A graphically simpler method of obtaining K_m is to locate the point where the extrapolated straight line crosses the abscissa, which gives the value $-(1/K_m)$. The similar triangles shaded in Fig. 20 show that the distance from the origin to the point marked $1/K_m$ is equal to the distance from the origin to the point marked $-(1/K_m)$. This conclusion may be stated algebraically by making $1/v$ equal to zero in the reciprocal equation, so that

$$0 = \frac{K_m}{v_{max}}.\frac{1}{[S]}+\frac{1}{v_{max}} \tag{48}$$

hence

$$\frac{1}{[S]} = -\frac{1}{K_m}. \tag{49}$$

Plot of v against $\dfrac{v}{[S]}$ [6, 50, 62]

In this method the basic equation (eqn. 27) is rearranged to the linear form:

$$v = v_{max}-\frac{v}{[S]}.K_m \tag{50}$$

This equation is derived as follows, starting again with equation (27) in its general form,

$$y=\frac{mx}{x+c}.$$

Multiplying through by $(x+c)$ and rearranging the equation gives

$$yx=mx-yc,$$

which is the same as

$$y=m-\frac{yc}{x},$$

and this is the general form of equation (50). Since m and c are constants, y is proportional to y/x.

Fig. 21 shows experimental data plotted in this way. v_{max} is the rate of transfer when $[S]=\infty$ and $(v/[S])=0$, so that v_{max} is that point at which the plotted line cuts the ordinate.

When v_{max} has been found, K_m is obtained by finding the point on the ordinate at which v is $\frac{1}{2}.v_{max}$ (A in Fig. 21). The corresponding point on the abscissa has the value

$$\frac{\frac{1}{2}.v_{max}}{K_m}$$

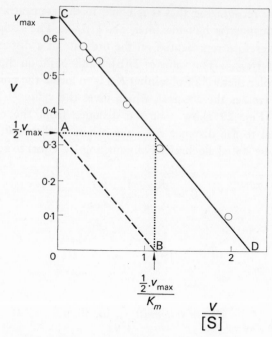

Fig. 21. Plot of v against $v/[S]$ for the saturable component of the uptake of benzylamine by Ehrlich ascites carcinoma cells.

$$K_m = \frac{A}{B} = \frac{\frac{1}{2} \cdot v_{max}}{\frac{1}{2} \cdot v_{max}/K_m} = \frac{C}{D} = 0\cdot3 \text{ mM}.$$

v = rate of uptake of benzylamine, mmoles/kg cell water/min. [S] = concentration of benzylamine in suspending medium, mM. Incubation 1 min, 37°C. (Redrawn from data published by Christensen & Liang [34].)

(B in Fig. 21) from which the value of K_m can be readily calculated. K_m can also be expressed as the ratio of the intercepts of the plotted line with the ordinate (C in Fig. 21) and abscissa (D in Fig. 21), i.e. C/D. As can be seen, this line is parallel with the line AB, so that A/B = C/D. Since

$$\frac{A}{B} = \frac{\frac{1}{2} \cdot v_{max}}{\frac{1}{2} \cdot v_{max}/K_m} = K_m,$$

C/D = K_m.

Plot of $\dfrac{[S]}{v}$ against [S] [50, 58, 62]

The kinetic equation (eqn. 27) is rearranged in the linear form:

$$\frac{[S]}{v} = \frac{1}{v_{max}} \cdot [S] + \frac{K_m}{v_{max}} \tag{51}$$

This equation can be derived by referring to the general form of equation 27,

$$y = \frac{mx}{x+c},$$

which is first converted to its reciprocal,

$$\frac{1}{y} = \frac{x+c}{mx}.$$

Multiplying through by x gives

$$\frac{x}{y} = \frac{1}{m}.x + \frac{c}{m}.$$

which is the general form of equation (51) .Since m and c are constants, x/y is proportional to x. The relationship is shown graphically in Fig. 25.

The slope of the straight line so derived is $1/v_{max}$, as can be seen from the equation, and from this the value of v_{max} may be calculated. If in the equation, $[S]/v$ is made equal to zero, then $[S] = -K_m$. Thus $-K_m$ can be read off the plot as the value of $[S]$ at which the line cuts the abscissa.

A property of this plot, not shared by other graphical methods, is that even if a

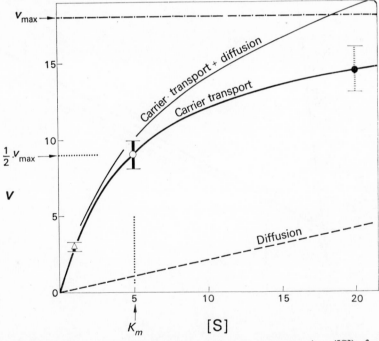

Fig. 22. Plot of rate of transfer (v) against substrate concentration ([S]), for carrier transport, diffusion and total transfer (carrier transport+diffusion), using hypothetical values in equations (9) and (27). The three *drawn symbols* have been inserted to enable direct comparisons to be made with Figs. 23, 24 and 25, which represent algebraical manipulations of the data used in the figure. The *vertical lines* associated with each symbol represent an error in v of ± 10 per cent. It is assumed that error in [S] is negligible. $v_{max} = 18$, $K_m = 5$, $K_D = 0\cdot2$; arbitrary units.

diffusion component is not first subtracted from total transfer, the resulting curvilinear graph still cuts the abscissa at $-K_m$ (Fig. 25). This can be shown algebraically. The equation (eqn. 28) which describes such total transfer is:

$$v = \frac{v_{max}.[S]}{[S]+K_m} + K_D.[S].$$

Dividing through by [S], gives:

$$\frac{v}{S} = \frac{v_{max}}{[S]+K_m} + K_D = \frac{K_D.[S] + K_D.K_m + v_{max}}{[S]+K_m} \tag{52}$$

which by inversion gives:

$$\frac{[S]}{v} = \frac{[S]+K_m}{K_D.[S] + K_D.K_m + v_{max}} \tag{53}$$

When $[S]/v = 0$.

$$\frac{[S]}{K_D.[S] + K_D.K_m + v_{max}} = -\frac{K_m}{K_D.[S] + K_D.K_m + v_{max}} \tag{54}$$

hence $[S] = -K_m$.

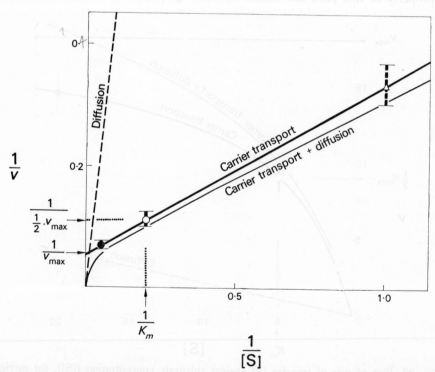

Fig. 23. Double reciprocal plot derived from the same hypothetical values as those used in Fig. 22 (cf. Fig. 20). The appearance of the 10 per cent error in v associated with each symbol (see legend to Fig. 22) has become distorted by manipulation of the data. Carrier transport and diffusion are both represented by straight lines, but the line for total transfer (carrier transport + diffusion) is curved and passes through the origin. Symbols as in Fig 22. $v_{max} = 18$, $K_m = 5$, $K_D = 0.2$; units arbitrary.

The relationships between the above types of plot are shown in Figs. 22, 23, 24 and 25. Fig. 22 shows a plot of the rate of carrier transport (v) against the concentration of substrate ([S]). Values of v at a low substrate concentration, at a high concentration and at an intermediate concentration (equal to the value of K_m) are each shown with an error of ± 10 per cent. Since error in the value of substrate concentration is usually rather small, none has been assigned to the values of [S] in the figures. In addition, lines corresponding to diffusion and to total transfer (diffusion + carrier transport) have been included. Fig. 23 shows the conversion of all data in Fig. 22 to a double reciprocal plot, Fig. 24 to a plot of v against $v/[S]$, and Fig. 25 to a plot of $[S]/v$ against $[S]$.

Evaluation of straight-line conversions

'One can't believe impossible things', said Alice. 'I daresay you haven't had much practice', said the Queen. 'When I was your age, I always did it for half-an-hour a day.

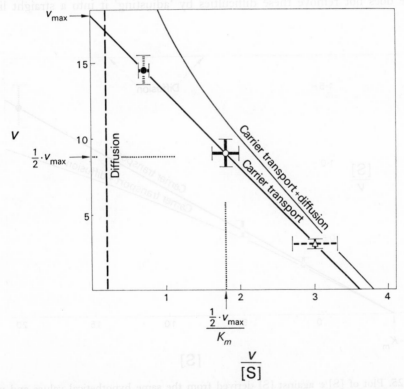

Fig. 24. Plot of v against $v/[S]$ derived from the same hypothetical values as those used in Fig. 22 (cf. Figs. 21, 37 and 38). The appearance of the 10 per cent error in v associated with each symbol (see legend to Fig. 22) has become distorted by manipulation of the values, and, since v is present in the scale of both ordinate and abscissa, here produces an error related to each scale. Carrier transport and diffusion are both represented by straight lines, but the line for total transfer (carrier transport + diffusion) is curved. Symbols as in Fig. 22. $v_{\max} = 18$, $K_m = 5$, $K_D (= (v/[S])) = 0.2$; units arbitrary.

Why, sometimes I've believed as many as six impossible things before breakfast . . .' [26]. This discourse could and sometimes does seem to apply to the assessment of kinetic data; the following is aimed at discouraging such habits.

The three algebraical rearrangements described above (eqns. 47, 50 and 51) say neither more nor less than the original kinetic equation, since they are simply the result of algebraical manipulations. The essential statement is in no way altered by them. If, owing to the extent of the scatter of experimental values, it is difficult to draw a curve through the points when plotting rate of transfer against concentration of substrate, it will be just as unsatisfactory to draw a straight line through values which have been manipulated as described. It is easy enough to do it, but the belief that the relationship of such a line to the experimental values is in some way any better is delusory; it is the validity of that line to those values and the justification (if any) in projecting it beyond the experimental values that is all-important.

For example, if the shape of the saturation curve which is drawn has certain inconveniences, especially in relation to one 'end' or the other (perhaps its asymptotic character), one does not remove these difficulties by 'adjusting' it into a straight line. The

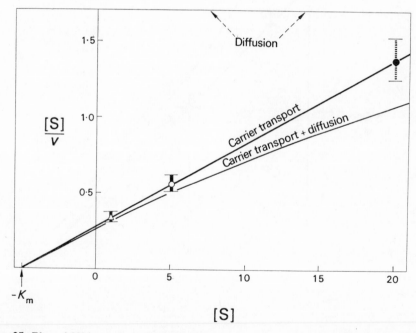

Fig. 25. Plot of [S]/v against [S] derived from the same hypothetical values and with the same symbols as used in Fig. 22. The appearance of the 10 per cent error in v associated in v associated with each symbol (see legend to Fig. 22) has become distorted by manipulation of the values; the distortion is the reverse of that in Fig. 23. Carrier transport and diffusion are both represented by straight lines (diffusion is represented by a horizontal line outside the limits of the diagram), but the line for total transfer (carrier transport+diffusion) is curved and cuts the abscissa at a value of [S] equal to $-K_m$ (see text). $K_m = 5$, $v_{max} = 18$, $K_D = 0.2$, $1/K_D (= [S]/v) = 5$; units arbitrary.

uncertainties about which curve to draw are *not* improved by an 'adjustment' (whose justification often implies an unspoken assumption) such as putting the data into some linear transformation. The advantage of such a procedure is that it is visually easier to recognize the properties of a straight line than those of a rectangular hyperbola. If experimental values do all lie close to a hyperbola which can be described by the Michaelis–Menton equation, a linear plot reveals that property, and from the linear plot the values of v_{max} and K_m may be extracted. In practice, experimental results which are sufficiently consistent and, at the same time, adequate in number appear to be uncommon in the field of membrane transfer, just as they do in the field of enzyme kinetics [51].

The double reciprocal plot is the most often used of the three straight-line conversions which we have described. It may, nevertheless, be the least reliable source of estimates of v_{max} and K_m, although there is some disagreement on this point. The three methods, as they are used in enzyme kinetics, have been compared by Dowd & Riggs [51] who analysed them by using simulated data, and investigated the sensitivity to error in the parameters. They found that the double reciprocal plot tends to give a deceptively good 'fit', even with unreliable experimental values. If used without adequate 'weighting'

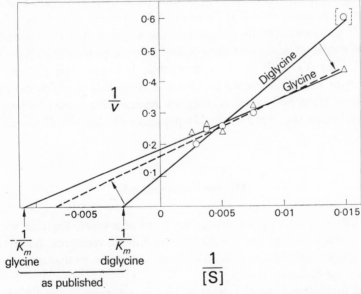

Fig. 26. The effect of a single value on the interpretation of a double reciprocal plot. Comparison of K_m of glycine ($\triangle$) and of diglycine ($\bigcirc$) during transfer by intestine. 'Best fit' lines through the two groups of experimental points are shown by solid lines. If the single experimental point for diglycine at the highest value of $1/[S]$ is omitted (square brackets), the 'best fit' straight line for diglycine is shown by the interrupted line; K_m for glycine would then be about 110 mM, i.e. not very different from that for diglycine, in contrast to the published values of 91 mM and 328 mM respectively, which are claimed to be significantly different at $P < 0.05$. v = rate of transfer (μmole/cm/5 min); $[S]$ = initial concentration of substrate indicated, mM. (Redrawn from data published by Matthews *et al.* [93].)

(i.e. without allowing for the undue influence exerted by single experimental points at high values of 1/[S]) it can result in large errors. An example of this is shown in Fig. 26, which contains the results of experiments carried out to compare the transfer of glycine with its peptide diglycine [93]. In each case five points were plotted and a straight line of best fit drawn by the original authors through each group (shown by the solid lines). These two lines diverge from each other so greatly that the authors concluded that the kinetic constants for the two substrates were notably different. However, if the single point in the top right-hand corner of Fig. 26 is omitted, the line of best fit through the remaining four points for diglycine is little different from the line representing glycine; this also shows the danger of inadequate experimental data.

Dowd & Riggs suggest, indeed, that the double reciprocal plot should be abandoned as a method for estimating kinetic constants from unweighted values. There is, in their opinion, little to choose between the other two methods, although plotting v against $v/[S]$ seems slightly preferable (especially if error in the measurement of v is likely to be large), since the properties of this plot (Fig. 24) tend to discourage false assumptions.

Preferably, transfer rates should be obtained over as wide a range of concentration as possible. In practice this is not usually done, since the use of very high concentrations of substrate may have serious difficulties, while at very low concentrations errors of estimation of the rate of transfer are likely to increase. It is important that the range of concentration used should extend as far as possible on either side of the value of K_m for the system. If concentrations lie only below the value of K_m, the estimation of v_{max} (and hence of course of K_m also) becomes unreliable, and if values above K_m only are used, the value for K_m itself becomes unreliable. It is revealing that of twenty-eight papers on enzyme kinetics surveyed by Dowd & Riggs [51] in which double reciprocal plots were used, eight had no measurements representing concentrations above K_m. The same deficiencies may also be found in papers on the kinetics of membrane transfer [14, 68, 70, 131].

Algebraical methods

Experimental data are often presented in the form of a mean, together with statistical estimate of accuracy such as the standard error. Kinetic constants, however, are often presented as a single value only, without indication of error, so that no estimate of their reliability can then be made.

A method for determining the error of kinetic constants derived from Michaelis–Menten kinetics has been described by Wilkinson [160]. It consists initially of the determination of approximate values of v_{max} and K_m, followed by the adjustment of these to a more accurate mean whose standard error is then found. Since the calculations are complicated the reader is referred to the original paper. The method has the advantage that values for kinetic constants can be obtained in an unbiased way, yet are weighted to allow for distortion resulting from error in different regions of the kinetic curve. It has the disadvantage that calculations are lengthy and require a desk calculator for accuracy. It is assumed moreover that the experimental values used conform to

Michaelis–Menten kinetics, so that any deviation from these kinetics would simply produce a larger standard error for the values of the apparent constants calculated.

We suggest here an alternative method which is simpler, both in its approach and in the amount of calculation required. The kinetic equation (eqn. 27),

$$v = \frac{v_{max} \cdot [S]}{[S] + K_m},$$

contains two unknown values, v_{max} and K_m, so that if the rate of transfer (v) is found at two different concentrations of substrate ([S]), two equations are obtained which can then be solved to provide values for the two constants. A number of pairs of values of v and [S] may be taken, as shown in Fig. 27, and several values for v_{max} and K_m will be obtained. The mean of each with its standard error may then be calculated. In order to avoid bias, one pair of values should be taken in the concentration range of substrate below the value of K_m and one above this value. (A rough kinetic curve of the data can be drawn in the initial stages from which the approximate location of K_m can be

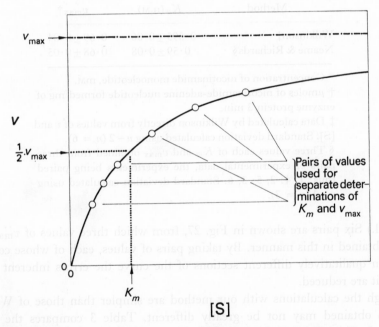

Fig. 27. Determination of kinetic constants by using pairs of experimental values on a single kinetic curve. If transfer follows Michaelis–Menten kinetics, K_m and v_{max} can be solved in the equation

$$v = \frac{v_{max} \cdot [S]}{[S] + K_m}$$

(eqn. 27) by obtaining two pairs of values for v and [S], and solving the resulting simultaneous equations. In each pair of values, one value of [S] should be above the value for K_m and one below. By taking a number of such pairs, several values for K_m and v_{max} are obtained which can then be analysed statistically (see Table 3). v = rate of transfer; [S] = concentration of substrate.

Table 3. Comparison of kinetic constants and standard error of the mean derived (i) by the method of Wilkinson [160] and (ii) by the method of the present authors as described in the text (and see Fig. 27). The same original data from six experiments carried out by Atkinson *et al.* [5] on enzyme action (for data see Wilkinson [160]) have been used in each case

Data used:

Experiment	[S]*	v†
1	0·138	0·148
2	0·220	0·171
3	0·291	0·234
4	0·560	0·324
5	0·766	0·390
6	1·460	0·493

Constants obtained (mean and standard error of the mean):

Method	K_m(mM)	v_{max}†
Wilkinson‡	0·595 ± 0·06	0·69 ± 0·04
Neame & Richards§	0·59 ± 0·08	0·68 ± 0·05

* Concentration of nicotinamide monocleotide, mM.
† μmoles of nicotinamide-adenine nucleotide formed/mg of enzyme protein/3 min.
‡ Data calculated by Wilkinson directly from values of v and [S]. Standard deviation calculated using $n-2$ ($n = 6$).
§ Three values each of K_m and v_{max} obtained from three pairs of experimental data, the experiments being paired thus: 1, 4: 2, 5: 3, 6. Standard deviation calculated using $n-1$ ($n = 3$).

determined.) Six pairs are shown in Fig. 27, from which three values of v_{max} and K_m could be obtained in this manner. By taking pairs of values, each of whose components comes from qualitatively different sections of the curve the errors inherent in special regions of it are reduced.

Although the calculations with our method are simpler than those of Wilkinson's, the results obtained may not be greatly different. Table 3 compares the results of calculations using the two methods on kinetic data derived from enzyme experiments.

Determination of diffusion constant in the presence of carrier transport

When the methods for the determination of the kinetic constants of one-way carrier transport were described, the possible presence of diffusion was not discussed, although, of course, diffusion of substrate will usually also occur. In some cases, the rate of diffusion may be so low as not to be detectable or to be unimportant within the framework of the particular experimental design. Often, however, the contribution of diffusion to the total rate of transfer may be so great that its magnitude must be known if the properties of a carrier are to be defined with accuracy. To determine this, diffusion may be isolated by elimination of carrier activity through chemical or physical means, or its contribution may be determined from total kinetic activity. Neither of these methods is entirely satisfactory.

The measurement of diffusion by elimination of carrier transport

If carrier transport is eliminated, the 'initial rate' of transfer by diffusion will be proportional to the concentration of solute, so that a plot of v against [S] will be linear, and the plotted line will pass through the origin and have a slope which represents the diffusion constant, i.e. $K_D = v/[S]$, from equation (9) (see Fig. 2).

Carrier transport can seldom be entirely eliminated with certainty, but can often be considerably reduced by the following methods:

(a) Temperature reduction

At temperatures near $0°C$ any carrier transport which primarily depends upon metabolic energy sources will be greatly reduced. Although carrier transport can sometimes appear to be eliminated in this way [15], this is by no means always the case [57], and so must not be assumed to be so.

(b) Metabolic inhibitors

Metabolic inhibitors may reduce the rate of carrier transport by interference with the source or transfer of energy at one or more points in the energy cycle [11, 95, 96, 106], but the energy block so produced may be incomplete. Consequently it can never properly be assumed that a metabolic inhibitor abolishes carrier transport entirely. An effective metabolic inhibitor may affect the integrity or the organization of a membrane across which transfer occurs and so may also indirectly alter diffusion characteristics.

(c) Changes in the ionic balance of a suspending medium

There is evidence that the presence of certain inorganic ions may be essential for effective carrier transport of metabolites across biological membranes [139a]. For example, if a sodium-free medium bathes a segment of intestine, carrier transport of amino acid appears to be abolished [123]. Sugar transfer is also greatly reduced, and this is associated with a large increase in the value of K_m [40]. This suggests competitive inhibition (see Chapter 5, p. 56) but is more probably due to removal of a sodium 'facilitation' [139a].

The measurement of diffusion by saturation of carrier

If diffusion and carrier transport occur together, they may be distinguished from each other by saturation of the carrier component. At a high concentration of substrate

E

relative to K_m the carrier will be virtually saturated; if the concentration of substrate is now raised progressively any further increase in transfer will be attributable almost entirely to diffusion, and from this the diffusion constant may be determined either graphically or algebraically.

(a) Graphical method

The relationship described above is shown in Fig. 28. The line drawn through experimental points ('carrier transport+diffusion') at high concentrations of substrate ([S]) will be almost straight, and virtually parallel to a line representing the component of diffusion. In these conditions total transfer consists of a virtually constant component, represented by v_{max}, together with a component of diffusion whose rate will vary directly with the concentration of substrate. The line representing total transfer would, if extrapolated backwards, cut the ordinate at a value equal to v_{max}, as shown in Fig. 28. In practice, the value so obtained will be an approximation only, since the carrier is unlikely to be totally saturated in the range of concentrations usually employed.

(b) Algebraical method

It has been claimed that the diffusion constant (and also v_{max} and K_m) may be found algebraically by the measurement of transfer rates over prolonged periods of time [2, 88].

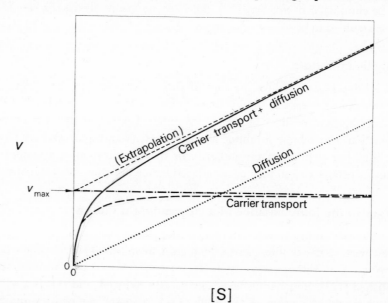

Fig. 28. Determination of diffusion constant (K_D) and maximal rate of transport (v_{max}) by the use of high concentrations of substrate. If the concentration of substrate is high enough the carrier will be approaching saturation and any excess transfer at still higher concentrations will be mainly due to diffusion. The line representing total transfer can then be extrapolated backwards as a straight line having a slope (dv/d [S]) approximately equal to the value of K_D (see Fig. 2) and cutting the ordinate at a value approximately equal to the value of v_{max}. v = rate of transfer; [S] = concentration of substrate.

The method rests on an assumption that inward transfer into a cell is by both carrier and diffusion, but that outward transfer is by diffusion only. The validity of this assumption must, in general, clearly be difficult to support. Nevertheless, within the limiting conditions above, the equation which represents the net rate of transfer inwards (V) at a specified time (t) is:

$$V = \frac{d[S']}{dt} = \frac{v_{max} \cdot [S]}{[S] + K_m} + K_D ([S] - [S']),$$

where d$[S']$ represents a very small change in intracellular concentration in a very small interval of time (dt). Integration of the equation produces the following result:

$$\frac{[S']}{[S]} = \frac{1}{K_D} \cdot \frac{v_{max} \cdot [S]}{[S] + K_m} (1 - \exp(-K_D t)) \tag{55}$$

If the concentration is high enough the carrier approaches saturation, and the expression

$$\frac{v_{max} \cdot [S]}{[S] + K_m}$$

becomes virtually a constant. Under these conditions a plot of $[S']/[S]$ against $1/[S]$ gives a straight line which intercepts the ordinate, $[S']/[S]$, at a value equal to $(1 - \exp(-K_D t))$, from which the value of K_D can be determined. This value is then inserted into the integrated equation, together with experimental values of net transfer obtained at lower concentrations, so giving a number of values of transfer rate for the carrier alone, i.e. values for the expression

$$\frac{v_{max} \cdot [S]}{[S] + K_m}.$$

Values for v_{max} and K_m may then be obtained by any of the methods described earlier in this chapter.

Althernatively, if the diffusion constant alone is required, and if the assumptions upon which this method is based are accepted, a more accurate value for it would be obtained by the measurement of the movement of substrate out of a 'loaded' cell. Since this movement is assumed to be the result of diffusion alone, the required information would presumably be thus obtained directly.

It can be seen that there are several methods of analysing and presenting experimental data, but it is important to appreciate that each can only be used for particular experimental conditions. A one-way transfer model, with its attendant algebra, must clearly not be applied to data which are not a record of this kind of transfer. Nor can they be applied, without modification, to transfer by more than one carrier at the same time. It will become apparent (Chapter 6, p. 80) that it is not always easy to decide whether more than one carrier is involved in transfer. In such cases, as well as in those described above, adequate and careful experimental replication is probably the most important single factor. In our experience, 'adequate' usually means at least 50 per cent more than one has.

Chapter 5

Inhibition

The rate of carrier transport of a substrate across a membrane may be altered by the presence of other solutes, either by interference with the rate of attachment of substrate to carrier sites, or by interference with the metabolism of the cell upon which transport may depend.

In the first type, known as 'competitive inhibition', the carrier sites may be receptive to both substrate and to the interfering solute, which itself may or may not be transported. Since the sites are limited in number, occupation of some of them by a second solute (or 'competitor') reduces the number of available to substrate and hence its rate of transport. Alternatively, in certain cases the presence of an interfering solute may distort the carrier sites so that their receptivity for substrate is altered. These alterations may make the sites more 'receptive' or less 'receptive', and it is with this last that we are presently concerned; the K_m of the substrate whose transfer is being studied then has the appearance of being increased as a result of the reduced availability of sites, but there is no effect on v_{max} [40, 159].

In the second type (known as 'non-competitive' or 'metabolic' inhibition), energy-dependent transfer can be reduced by the presence of metabolic inhibitors such as cyanide, 2,4-dinitrophenol or iodoacetate. There is believed to be no effect on the attachement of substrate to carrier sites, although it would be unwise to assume that this is always so. For example, non-competitive inhibitors of enzymes may act by reducing the number of effective sites on those enzymes [50], and there may be a parallel to this in carrier transport; alternatively, the disturbance of energy transformation may influence the structure of the membrane. The value of v_{max} is reduced, and hence the rate of transport is reduced proportionately at all concentrations of substrate [40, 159].

The two types of inhibition are usually distinguished experimentally by determining the kinetic constants of a substrate in the presence and absence of inhibitor.

Competitive inhibition

It is easier to view competitive inhibition of a carrier initially in terms of the simpler concept of an adsorption surface, after which it can be modified to apply to transfer. The kinetics will be developed from the theory described in Chapter 3 (p. 16).

Let us assume that two substrates can be adsorbed on to the same sites. Their concentrations may be altered relatively to each other, and the effects of these changes upon adsorption of either may be observed. The probability that a site will be occupied by either substrate depends on two things, (i) the relative concentrations of the substrates and hence the likelihood that one species of molecule rather than the other will be in a particular place at a particular time, and (ii) their relative abilities (reflected by the value of each K_m) to remain adsorbed on to sites.

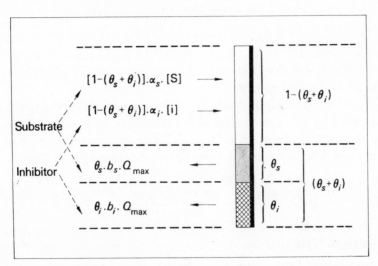

Fig. 29. Kinetic relationships of competitive inhibition on an adsorption surface at equilibrium. 'Substrate' and 'inhibitor' correspond to molecules of two different species either of which can be adsorbed on to sites on the surface, and both of which are treated in the same way kinetically. The *broad vertical line* represents the adsorption surface, the *adjacent shaded rectangle* the area occupied by 'substrate' molecules, and the *cross-hatched rectangle* the area occupied by 'inhibitor' molecules. Although represented by rectangles for convenience, occupied sites are randomly distributed over the whole adsorption surface. For interpretation of symbols see text and Fig. 9.

Essentially each substrate inhibits the other at all times but, by convention, one of the substrates (say x) is selected and its behaviour in the presence of the other (say y) is observed. x is frequently referred to as 'substrate' or 'S' and y as the 'inhibitor' or 'i'. It is neither more nor less reasonable to study the behaviour of y in the presence of x, i.e. of 'inhibitor' in the presence of 'substrate', for each has its own kinetic relationships with the surface, that is, its own association and dissociation constants. These constants (pp. 20 and 21) may be referred to as 'α_s' and 'b_s' in the one case and 'α_i' and 'b_i' in the other. At equilibrium each occupies a fraction of the adsorption surface, that is, 'θ_s' and 'θ_i' respectively. The fraction of unoccupied surface must then be $[1-(\theta_s+\theta_i)]$, and this is the 'area' available for the further association of either of the substrates with increase of concentration (Fig. 29).

The rate of association of one substrate will be $[1-(\theta_s+\theta_i)].\alpha_s.[S]$ and of the other $[1-(\theta_s+\theta_i)].\alpha_i.[i]$ and the rate of dissociation $\theta_s.b_s.Q_{max_s}$ and $\theta_i.b_i.Q_{max_i}$ respectively, where [S] and [i] are the concentrations of substrate and inhibitor, and Q_{max_s} and Q_{max_i} the respective masses adsorbed when saturated by each (cf. adsorption, p. 20). The equilibrium state of each can then be expressed as (cf. equation (16)):

For 'S':

$$[1-(\theta_s+\theta_i)].\alpha_s.[S] = \theta_s.b_s.Q_{max_s} \tag{56}$$

For 'i':

$$[1-(\theta_s+\theta_i)].\alpha_i.[i] = \theta_i.b_i.Q_{max_i} \tag{57}$$
$$\text{(rate of association)} \quad \text{(rate of dissociation)}$$

This is, in essence, all there is to competitive inhibition on an adsorption surface; everything else which follows is simple algebraical rearrangement which may make certain things clearer, but does not and cannot reveal anything which has not already been said in the two equations above.

The object of further manipulation is to obtain an equation describing the proportion of surface occupied by only one or other of the substrates, i.e. θ_s or θ_i, from which the behaviour of each can be determined in terms either of adsorption, or, when adapted, of carrier transport. The first equation (eqn. 56) can be rearranged to give:

$$\theta_s = \frac{\alpha_s.[S]-\alpha_s.[S].\theta_i}{\alpha_s.[S]+b_s.Q_{max_s}} = \frac{[S].(1-\theta_i)}{[S]+\dfrac{b_s.Q_{max_s}}{\alpha_s}} = \frac{[S].(1-\theta_i)}{[S]+K_s} \tag{58, 59, 60}$$

and the second (eqn. 57) to give:

$$\theta_i = \frac{\alpha_i.[i]-\alpha_i.[i].\theta_s}{\alpha_i.[i]+b_i.Q_{max_i}} = \frac{[i].(1-\theta_s)}{[i]+\dfrac{b_i.Q_{max_i}}{\alpha_i}} = \frac{[i].(1-\theta_s)}{[i]+k_i} \tag{61, 62, 63}$$

where 'K_s' and 'k_i' are the equilibrium constants of the two substrates.

If the expression for θ_i (eqn. 63) is substituted in equation (60) and this equation rearranged, θ_s can be obtained in relation to [S] and [i], thus:

$$\theta_s = \frac{[S].\left[1-\dfrac{[i].(1-\theta_s)}{[i]+k_i}\right]}{[S]+K_s}. \tag{64}$$

This can then be rearranged so that θ_s is expressed in terms of the other constituents:

$$\theta_s = \frac{[S].k_i}{[S].k_i+k_i.K_s+[i].K_s} \tag{65}$$

$$= \frac{[S]}{[S]+K_s+\dfrac{K_s.[i]}{k_i}} \tag{66}$$

$$= \frac{[S]}{[S]+K_s\left(1+\dfrac{[i]}{k_i}\right)}. \tag{67}$$

By following the same procedure, but substituting θ_s for θ_i the proportion of surface occupied by the other substrate, i, can be shown, thus:

$$\theta_i = \frac{[i]}{[i]+k_i\left(1+\frac{[S]}{K_s}\right)},$$

(68)

which, of course, has the same basic structure as equation (67).

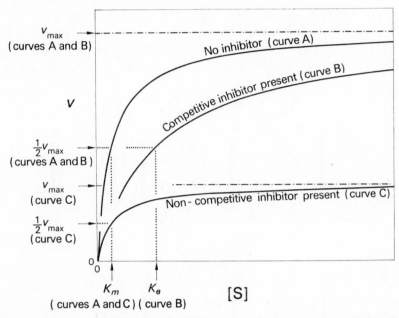

Fig. 30. Effect of a competitive and a non-competitive inhibitor on K_m and v_{max}. *Curve A:* the transfer of substrate in the absence of inhibitor. *Curve B:* the transfer of substrate in the presence of a competitive inhibitor; K_m appears altered to a value equal to K_a, where

$$K_a = K_m\left(1+\frac{[i]}{K_i}\right)$$

but v_{max} is unaltered. *Curve C:* the transfer of substrate in the presence of a non-competitive inhibitor; the value of v_{max} is reduced, but that of K_m is unchanged. v = rate of transfer; [S] = concentration of substrate.

If the substrate is transported, the rate of transport can be found by repeating the procedure described in Chapter 3, since the 'forward' dissociation constant of substrate, i.e. γ, is assumed to be unaffected by the presence of inhibitor. The rate of transport, v, may then be expressed as shown in equation (26) (i.e. $v = \theta . v_{max}$). At the same time, if both substrate and inhibitor are transported, the Michaelis constant replaces the dissociation constant for each respectively, the Michaelis constant of the inhibitor being designated K_i. The equation describing the rate of transport of substrate in the presence of inhibitor is then:

$$v = \frac{v_{\max} \cdot [S]}{[S] + K_m \left(1 + \dfrac{[i]}{K_i}\right)}.$$

(69)

If it should happen that the inhibitor is not transported, the Michaelis constant of the inhibitor would have to be replaced by its dissociation constant, that it, K_i would be replaced by k_i.

This equation is identical in form with equation (27), except that K_m has now been replaced by an 'Apparent K_m', i.e. $K_m(1 + ([i]/K_i))$, or, more briefly, K_a.

K_m thus appears to be increased in value by a factor $[i]/K_i$, as shown in Fig. 30 (cf. Figs. 10 and 11). For a particular value of the ratio $[i]/K_i$, and hence of K_a, the kinetic curve for the transport of substrate is altered, for example, from A to B, and at all finite

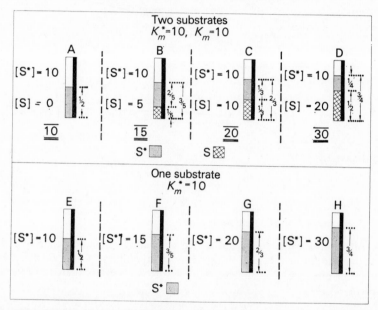

Fig. 31. The occupation of carrier sites by a labelled substrate in the presence and absence of its unlabelled form. In each case the *bold vertical line* represents the membrane and the *adjacent shaded rectangles* the proportion of sites occupied by each type of molecule, which would be distributed randomly over the whole surface (cf. Fig. 29). Units are arbitrary and the fractions to the right of each diagram represent the proportion of sites occupied. *Upper:* labelled (S*) and unlabelled (S) forms of the same substrate are present together, the labelled form at a single concentration throughout. As the concentration of the unlabelled form is increased (from *left* to *right*) the amount of labelled substrate which occupies sites decreases. *Lower:* labelled substrate present on its own, at a concentration in each case equal to the sum of the concentrations of the two forms in the diagram immediately above; in each case it occupies the same area. (The unlabelled form of substrate acts as a competitive inhibitor of the labelled form and vice versa, but here substrate and inhibitor have the same K_m; if each K_m were different, these relationships would not apply.) Calculated from equations (27) and (69).

concentrations of substrate ([S]) the rate of transport (v) is reduced. It should be clear that the kinetic properties would remain constant if either (i) the concentration of a particular inhibitor was unchanged, or (ii) K_i and [i] varied in proportion to each other.

As the concentration of inhibitor increases, the fraction of sites which it occupies also increases, leaving a smaller proportion of sites available to substrate. This is illustrated in Fig. 31, which shows the occupation of sites by radioactively labelled (S*) and unlabelled (S) forms of the same substrate. (The bold vertical lines in the figure represent membrane surface, and the shaded areas correspond to the number of sites occupied by each type of molecule. Although illustrated by block diagram, molecules are randomly distributed.) The labelled and unlabelled molecules are assumed to have the same K_m, and hence the ratio of the number of sites occupied by each form, i.e. θ_s^*/θ_s, will be the same as the ratio of their concentrations in the solution. This can be seen if K_m is substituted for K_i in the appropriate equation (comparable to eqns. 67 and 68) so that

$$\theta_s = \frac{[S]}{[S]+K_m\left(1+\dfrac{[i]}{K_m}\right)} = \frac{[S]}{[S]+K_m+[i]} \tag{70}$$

and

$$\theta_i = \frac{[i]}{[i]+K_m\left(1+\dfrac{[S]}{K_m}\right)} = \frac{[i]}{[i]+K_m+[S]} \tag{71}$$

from which

$$\frac{\theta_s}{\theta_i} = \frac{[S]}{[i]}, \text{ i.e. } \frac{\theta_s^*}{\theta_s} = \frac{[S^*]}{[S]}. \tag{72}$$

In the upper part of Fig. 31 the concentration of the labelled form is the same in all diagrams, but that of the unlabelled form varies. The reduced number of sites occupied by the labelled form in the presence of the unlabelled form (Fig. 31, B–D) gives the appearance of an increased value for the K_m of the labelled form, and if it were not known that the unlabelled form were present, it would indeed be measured as such (cf. Fig. 32, *lower*).

The diagrams in the lower half of the figure show for comparison the sites occupied by one substrate on its own at the total concentration shown in the corresponding upper sketch; the two representative areas are of course equal.

The effect of different values of the inhibitor constant (K_i) is shown in the upper portion of Fig. 32. In this case the concentrations of substrate and inhibitor are equal and the K_m of the substrate is unchanged, but that of the inhibitor (K_i) varies. In Fig. 32 (B) the constants of both substrate and inhibitor are the same, so that both occupy equal proportions of the sites available (as in Fig. 31, C). If the value of the K_m of the inhibitor (K_i) is smaller than that of the substrate (Fig. 32, A) the inhibitor is able to occupy a greater proportion of the sites at this concentration. Alternatively, if it is greater, the proportion is reduced (Fig. 32, C). The lower diagram shows for comparison the proportion of sites occupied by a single substrate having a K_m equal in value to that of the inhibitor in the upper diagrams (K_i). In Fig. 32 (F) the substrate is present at a concentration equal to its K_m and hence occupies half the number of sites, a situation similar to that in Fig. 32 (D).

The inhibitor equation (eqn. 69) (or some equivalent rearrangement of it) appears to have experimental support [29, 69, 76, 159]. It is identical in form with that used to describe competitive inhibition in enzyme kinetics [50] and is often used to interpret experimental data [43, 63, 100, 139]. Certain of its properties will now be examined.

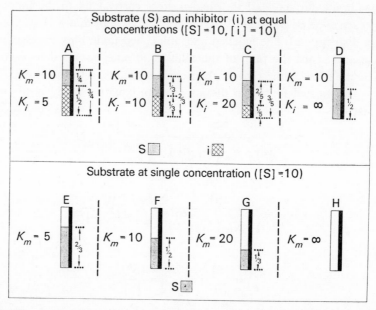

Fig. 32. The occupation of carrier sites by two substrates ('substrate' and 'inhibitor') at the same concentration. The general presentation is as shown in Fig. 31. *Upper:* K_m of substrate is the same throughout; K_m of inhibitor (i.e. K_i) varies. Inhibitor occupies fewer sites as K_i increases, while substrate occupies more. *Lower:* one substrate alone is present, whose K_m has the same value as K_i in the diagram immediately above. Calculated from equations (27) and (69).

Firstly, from fundamental assumptions, the value of the experimentally determined K_i of a substrate acting as an inhibitor should be the same as the value of the K_m of that substrate when transported alone. A comparison of this sort has been made for the entry of amino acids into Ehrlich ascites carcinoma cells and is shown in Table 4. In most instances the two values appear to be satisfactorily similar, but unfortunately no indication of the possible error of each is available. Another series along similar lines has also been carried out using brain slices, and is shown in Table 5. If the assumption is made that the errors here were in the range customary in this field of work, the values in Table 5 for K_m and K_i that would be considered in satisfactory agreement in each case are printed in heavy type, and hence could be regarded as support for the kinetic model described above. Some values are clearly not in agreement, indicating that the simple model does not always apply; these differences have been interpreted as evidence for transport by more than one carrier [16] (see Chapter 6, p. 80).

Table 4. Comparison between K_m and K_i for entry of amino acids and related compounds into the Ehrlich ascites carcinoma cell [37]. K_i was determined in the presence of the compounds indicated on the right. No indication of the possible error of these values was provided

Amino acid investigated	K_m(mM)	K_i(mM)	Substrate used for determination of K_i
MeAIB*	0·4	0·2	D-alanine, L-alanine
		0·3	Sarcosine, N-methylphenylalanine
		0·4	Proline
L-phenylalanine†	0·4	0·5	L-leucine, L-valine
L-leucine	0·6	0·5	L-valine
L-alanine	0·6–0·8	0·7	D-alanine
		0·9	MeAIB*
L-methionine	0·6–0·8	0·4	L-leucine
		0·6	D-valine
α-aminoisobutyrate	1·0	0·4, 0·7	L-alanine
L-proline	1·5	1·7	MeAIB*
		1·8	Glycine
Glycine	3–5	3·3	D-alanine
		3·4–4	α-aminoisobutyrate
Sarcosine	5	4	MeAIB*

* MeAIB = α-(methylamino)-isobutyrate.
† Phenylalanine has a K_i of about 11 mM when inhibiting transfer of α-aminoisobutyrate or α,γ-diaminobutyrate, which suggests the presence of more than one carrier system [34] (see Chapter 6, p. 80).

Table 5. Comparison between K_m and K_i for entry of amino acids into brain slices [16]. No indication of the possible error of the values was provided, but figures in heavy type represent values for K_m and K_i considered in each case to be in satisfactory agreement

Amino acid	K_m	K_i when acting as inhibitor of			
		AIB	L-phe	L-arg	L-asp
Glycine	**1·1**	**1·7**	7·5	21	11
L-alanine	**1·0**	**1·2**	3·2	23	
D-alanine	**2·4**	**3·0**	5·4		
1-amino-cyclopentane-1-carboxylate	**1·6**	**1·6**	**1·9**		
L-leucine	0·5	16	2·0	93	
L-methionine	0·7	6·4	2·0	13	
L-histidine	1·2	8·3	**1·8**		
L-lysine	**1·0**			**1·4**	
L-glutamate	0·5				2·0
D-glutamate	**3·0**				**1·8**

AIB = α-aminoisobutyrate
L-phe = L-phenylalanine
L-arg = L-arginine
L-asp = L-aspartate

Secondly, when the concentration of substrate is not changed, but that of inhibitor is progressively increased, the rate of transfer of the substrate should fall in a predictable manner, according to equation (69). Fig. 33 shows an example of this, in which increasing concentrations of the amino acid L-isoleucine progressively reduced the uptake of L-histidine by the intestine. The experimental values closely correspond to values calculated by inserting into the equation the kinetic constants derived from other experiments.

Thirdly, the effect of an inhibitor should depend on the relationship between the K_m of the substrate and the K_m of the inhibitor (i.e. its K_i). The lower the value of the K_m of the inhibitor relative to that of the substrate the greater should be its inhibitory

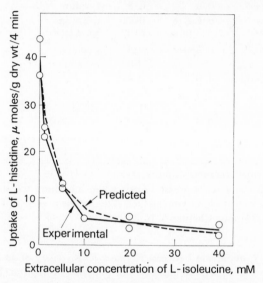

Fig. 33. Kinetics of competitive inhibition. Effect of progressive increase in the concentration of the amino acid L-isoleucine, as competitive inhibitor, on the rate of transfer of L-histidine at fixed extracellular concentration (10 mM) in segments of small intestine of the rat. Incubation 4 min. *Interrupted line* represents values predicted by the original authors from the Michaelis constants and the average uptake of L-histidine in the absence of L-isoleucine (L-histidine: $K_m = 10\cdot4$ mM; L-isoleucine: $K_m = 1\cdot2$ mM). (Redrawn from data published by Finch & Hird [56].)

effect. On the other hand, if the value of the K_m of the inhibitor is high enough, its effect on the transport of substrate will not be detectable. This can be illustrated by the entry of certain sugars into the erythrocyte in the presence of equimolar concentrations of other sugars (Table 6). A sugar (such as dextrose, i.e. D-glucose) with a relatively low K_m markedly reduced, when at the same concentration, the rate of entry of other sugars (such as sorbose or laevulose) with a relatively high K_m, although the opposite was not the case [78]. It is simple to calculate the transfer rate which may be expected in each case if those values for K_m shown in Table 6 are substituted in equations (27) and (69). Calculated values for the transport of sorbose and dextrose in the presence of each other at equimolar concentrations are shown in Table 7 (i) and (ii), and it will be seen that the numerical values of percentage inhibition correspond to the qualitative results in

Table 6. Mutual inhibition of uptake of sugars by the erythrocyte from solutions containing equimolar concentrations (150 mM) [78]

Substrate	K_m(mM) [76]	Inhibitor			
		Dextrose	Galactose	Sorbose	Laevulose
Dextrose	7–10	—	+ +	○	○
Galactose	50	+ + +	—	○	○
Sorbose	1300–2000	+ + + +	+ + +	—	○
Laevulose	*ca.* 2000	+ + + +	+ + + +	+ + +	—

+ + + + 'Essentially complete block of uptake' of substrate.
+ + + 'Very marked inhibition' of uptake of substrate.
+ + 'Moderate inhibition' of uptake of substrate.
○ 'No, or doubtful, effect' on uptake of substrate.

Table 6. It is also clear that while example (ii) in Table 7 suggests that inhibition of dextrose transport would not be detectable in these particular experiments, it is clear from example (iii) that appropriate adjustment of concentrations would probably reveal it. Confusion can obviously arise from concern with the concentration of the inhibitor rather than with the relative concentrations of inhibitor and substrate.

There are instances in which mutual inhibition is not readily demonstrable between competing substrates. For example, the amino acid methionine has a marked inhibitory effect on the transfer of glycine in the intestine, whereas glycine appears to have little or no effect on the movement of methionine [111, 164]. In such cases, where inhibition has not been examined over a wide range of concentrations it would be easy to conclude that one of the substrates had an inhibitory effect peculiar to itself.

Lastly, if the descriptive equation is valid, the presence of a competitive inhibitor, although apparently altering the value of K_m, should have no effect on the v_{max} of a

Table 7. Calculated rates of uptake of dextrose and sorbose by the erythrocyte in the presence and absence of each other. Rates found by inserting values for K_m, K_i (see Table 6), [S] and [i] into equations (27) and (69) and expressed as a percentage of v_{max}. (It is assumed that v_{max} is the same for each sugar)

Substrate	Substrate concentration (mM)	Transfer rate of substrate alone	Inhibitor	Inhibitor concentration (mM)	Transfer rate of substrate in presence of inhibitor	Inhibition
(i) Sorbose	150	$8 \cdot 1\% \, v_{max}$	Dextrose	150	$0 \cdot 5\% \, v_{max}$	93·8%
(ii) Dextrose	150	$94 \cdot 6\% \, v_{max}$	Sorbose	150	$94 \cdot 1\% \, v_{max}$	0·4%
(iii) Dextrose	1·5	$15 \cdot 0\% \, v_{max}$ {	Sorbose	150	$13 \cdot 9\% \, v_{max}$	7·3%
			Sorbose	1500	$8 \cdot 5\% \, v_{max}$	43·3%

Sorbose K_m = 1650 mM.
Dextrose K_m = 8·5 mM.

substrate. This has been verified experimentally in conditions in which the structure of substrate and inhibitor were such that there could be no reasonable doubt that the inhibition observed was competitive, as for example would be the case with stereoisomers [3, 16, 40, 113, 143].

In summary, it seems that in many cases the model and its associated equations are borne out by experimental data. If, however, after rigorous test, cases of 'deviation' remain, a modification of the simple competitive inhibitory model must be made.

Inhibition by high concentration of substrate

It is sometimes found in enzyme kinetics that at low concentrations of substrate a reaction rate follows Michaelis–Menten kinetics, but at higher concentrations it is reduced ('inhibition by excess substrate') [50]. The phenomenon has been attributed to alterations in the attachment of substrate molecules to enzyme. When a substrate molecule has to be attached at two or more group-specific points on the enzyme for effective activity, over-crowding of substrate molecules at high concentrations may reduce the number of points at which each molecule is attached, thereby reducing the effectiveness of the enzyme [50].

This may also occur during the transfer of substrates across membranes, since there is, for example, some evidence, albeit inconclusive, of a reduction in the rate of transfer of certain amino acids at high concentrations [56, 94]. Amino acids probably have to be attached to carrier by three points for optimal transport [28, 29, 35], and changes in the three-dimensional arrangement of these points may reduce the rate of transfer [106, 107]. This suggests that at those sites where there is suboptimal attachment associated with high substrate concentrations, there will be partial inhibition of transport.

Algebraically this is treated like competitive inhibition. If we consider it first of all as adsorption, a proportion, θ_1, of sites is occupied by single molecules of substrate and a proportion, θ_2, by, say, two molecules, which leaves a proportion, $1-(\theta_1+\theta_2)$, unoccupied.

There are then two equilibria for substrate. One is of the type already described (eqn. 56) and related to the one-molecule one-site attachment, thus:

$$[1-(\theta_1+\theta_2)].\alpha_1.[S] = \theta_1.b_1.Q_{max_1} \qquad (73)$$
$$\text{(rate of association)} \qquad \text{(rate of dissociation)}$$

The second equilibrium is that involving attachment sites holding one molecule which then gain a second molecule:

$$\theta_1.\alpha_2.[S] \qquad = \qquad \theta_2.b_2.Q_{max_2} \qquad (74)$$

(association of one-molecule site with second molecule)　　　(dissociation of molecule from two-molecule site)

In each equation the subscript '1' designates a relationship between substrate molecule and site and '₂' a relationship between substrate molecule and substrate-site complex respectively. Equation (74) can be rearranged to:

$$\theta_2 = \frac{\theta_1 . \alpha_2 . [S]}{b_2 . Q_{\max_2}} \tag{75}$$

The function describing θ_2 is then substituted in equation (73). When multiplied out, this then appears as:

$$\alpha_1 . [S] - \theta_1 . \alpha_1 . [S] - \frac{\theta_1 . \alpha_2 . [S] . \alpha_1 . [S]}{b_2 . Q_{\max_2}} = \theta_1 . b_1 . Q_{\max_1} \tag{76}$$

This is rearranged to:

$$\theta_1 = \frac{[S]}{[S] + \dfrac{b_1 . Q_{\max_1}}{\alpha_1} + \dfrac{\alpha_2 . [S]^2}{b_2 . Q_{\max_2}}}, \tag{77}$$

and then simplified to:

$$\theta_1 = \frac{[S]}{[S] + K_{s_1} + \dfrac{[S]^2}{K_{s_2}}} = \frac{[S]}{[S] + K_{s_1}\left(1 + \dfrac{[S]^2}{K_{s_1} . K_{s_2}}\right)} \tag{78, 79}$$

where K_{s_2} is the equilibrium constant for the attachment of an extra molecule of substrate to each site already occupied by one substrate molecule. It is clear that the value of θ_1 is reduced when values of [S] become very high.

This can now be modified to express transport as described in Chapter 3 (eqn. 26) so that the rate of transport, v_1, which is by way of sites to which substrate molecules are optimally attached, can be expressed as:

$$v_1 = \frac{v_{\max_1} . [S]}{[S] + K_{m_1} + \dfrac{[S]^2}{K_{m_2}}} \tag{80}$$

To determine the rate of transport, v_2, by way of sites occupied by two molecules, equation (74) is rearranged to:

$$\theta_1 = \frac{\theta_2 . b_2 . Q_{\max_2}}{\alpha_2 . [S]} \tag{81}$$

which can be simplified to

$$\theta_1 = \frac{\theta_2 . K_{s_2}}{[S]} \tag{82}$$

This expression for θ_1 is substituted in equation (78), and the equation rearranged to give

$$\theta_2 = \frac{[S]^2}{[S]^2 + K_{s_1} . K_{s_2} + [S] . K_{s_2}} \tag{83}$$

This can then be modified to apply to transport (see equation (26)), thus:

$$v_2 = \frac{v_{\max_2} . [S]^2}{[S]^2 + K_{m_2}([S] + K_{m_1})} \tag{84}$$

The total rate of transport, v_{total}, is the combined rates of transport by way of each type (eqn. 80) and (eqn. 84), thus:

$$v_{\text{total}} = v_1 + v_2.$$

Competitive inhibition in two-way transfer

So far, competitive inhibition has been discussed in terms of one-way movement across a membrane. As already pointed out, transfer will in practice involve simultaneous movement in opposite directions; similarly, an inhibitor may affect outward transport at the same time as inward transport. Although there is evidence that related substances can interfere with each other's outward movement [81], data are insufficient at present to justify quantitative application of the kinetic model hitherto discussed.

If the inhibitor is adsorbed on to attachment sites on a carrier without itself being transferred into a cell, it will interfere only with *entry* of substrate into the cell. The equation for net entry (V) of substrate (omitting diffusion) would then be:

$$V = \frac{v_{\text{max}} \cdot [S]}{[S] + K_m \left(1 + \frac{[i]}{ki}\right)} - \frac{v_{\text{max}'} \cdot [S']}{[S'] + K_{m'}}. \tag{85}$$

(inward transport)   (outward transport)

If, however, the inhibitor is itself transferred into the cell, it will compete with substrate for both *inward and outward* transport sites, but the parameters for outward transport need not be the same as those for inward transport. The equation for net entry of substrate into the cell would then be:

$$V = \frac{v_{\text{max}} \cdot [S]}{[S] + K_m \left(1 + \frac{[i]}{K_i}\right)} - \frac{v_{\text{max}'} \cdot [S']}{[S'] + K_{m'} \left(1 + \frac{[i']}{K_i'}\right)} \tag{86}$$

(inward transport)   (outward transport)

Here, [i'] represents the concentration of inhibitor inside the cell, and $K_{i'}$ the K_m of the inhibitor for outward transport.

The interactions here are far more complicated than in the absence of inhibitor. There are not only the kinetic values for inward and outward movement of the substrate to be considered, but also those for inward and outward movement of the inhibitor. With the passage of time, not only will the intracellular concentration of substrate and of inhibitor change, but the rates of these changes may well be different.

It has already been pointed out that in 'initial rate' experiments, a substantial proportionate error can arise from the simultaneous exit of substrate from a cell (e.g. Table 2). In 'initial rate' experiments in which inhibitors are transferred into the cell it becomes even more necessary to recognize the importance of error in calculated kinetic parameters.

Activation

The behaviour which, in enzyme kinetics, is termed 'activation' [50] seems to occur also in carrier transport. It is the reverse of competitive inhibition in the sense that the presence of increasing concentrations of an added solute, here called an 'activator', increases the rate of reaction. Activators of enzyme reactions are usually metal ions which increase the reactivity of adsorption sites on an enzyme [50], and correspond to the inorganic catalysts of non-biological chemistry.

There is as yet only limited evidence of the existence of activators in biological transfer systems. As an example, the extracellular concentration of sodium ions can affect the rate of transfer of a substrate [10, 36, 43, 92, 139a], and this may be associated with an apparent alteration in the value of K_m [70, 89, 139a]; for instance, the uptake of 6-deoxyglucose by intestine is apparently increased by increasing the concentration of Na^+ [41]. Alternatively, certain intracellular amino acids can increase the rate of exit of others in brain tissue [81], but in this case precise kinetic relationships have not been determined.

A substance could behave as an activator if its presence on a carrier so altered the receptor sites for substrate that the latter 'fitted' on to the carrier more readily [40] (roughly, activator would then be acting as a 'co-carrier' by analogy with co-enzyme). Increasing concentrations of activator would increase the number of sites affected and so reduce the value of K_m [40], but the total number of sites (and hence v_{max}) would be unchanged. Activator itself might be transported, so that a triple complex was formed of which two components were transported [43].

Alternatively, activator could in theory be transported at sites not normally transporting substrate, but if substrate could become attached to activator, it would then acquire a second route of entry. The presence of activator would then in effect increase the number of sites available, and hence increase v_{max}.

Non-competitive inhibition

In this type of inhibition, attachment of substrate to available sites is believed to be unimpeded by the presence of inhibitor at all concentrations of substrate. The inhibitor may act either by decreasing the turnover rate of carrier sites, or by decreasing the amount of carrier available, or both. Experimentally, an inhibitor of this sort typically reduces v_{max} but does not affect the value of K_m [144].

The most usual type of non-competitive inhibitor is one which interferes with the production or utilization of energy in the cell, as with cyanide, 2,4-dinitrophenol [19, 46, 120] or ouabain [19, 121]. In isolated tissues the absence of metabolic substrate such as glucose from the environment may also inhibit carrier transport [96, 106], presumably by interfering with the production of oxidation-dependent energy. It follows that any agent which prevents access of metabolic substrates to intracellular sources of energy may also produce the same effect. This mode of inhibition (also known as 'metabolic'

F

inhibition) is most readily seen in concentrative transfer as opposed to equalizing transfer, possibly as a result of interference with changes in carrier configuration which are believed to occur in this type of transfer.

The mechanism of non-competitive inhibition in carrier transport presumably differs from that in enzyme kinetics where the inhibitor forms a complex with substrate and sites and in which a constant, K_i, can be derived in a manner similar to that of competitive inhibition, with a value equal to that concentration of inhibitor which reduces v_{max} by half. Since the mechanism of action in carrier transport is probably different from that in enzyme kinetics, it would be unreasonable to derive a constant of this sort, although a concentration of inhibitor which reduces v_{max} by half may still be used as a point of reference.

ANALYSIS OF THE KINETICS OF COMPETITIVE INHIBITION (DETERMINATION OF K_i)

There are several methods of analysing the kinetics of competitive inhibition and hence of determining the value of K_i; some have been briefly described elsewhere in terms of enzyme kinetics [50]. Most are modifications of the methods used for finding the value of K_m (described in Chapter 4) and involve manipulation of the inhibition equation (eqn. 69) to obtain a straight-line plot on a graph. They have all the risks associated with such a plot (see Chapter 4). In some of the methods, the values of K_m and v_{max} of substrate in the absence of inhibitor are found as part of the procedure. It will be evident that the value obtained for K_i in terms of competitive inhibition has meaning only if competition is known to be wholly competitive, and if carrier transport in principle applies.

For an accurate estimation of the value of K_i it is best that inhibitor and substrate are used at concentrations in the region of the value of K_i and K_a respectively, since in each case kinetic behaviour will then be changing most rapidly with change in concentration.

In the following description of some methods it will be assumed that any component of diffusion is negligible, or that it has first been substrated from the values measured.

Graphical methods

VARIATION OF CONCENTRATION OF SUBSTRATE WITH FIXED CONCENTRATION OF INHIBITOR

Double reciprocal plot [50]

The procedure is identical with that used for determining K_m and v_{max} (p. 41), except that two sets of values are obtained, one for the transfer of substrate in the absence of inhibitor and one for the transfer of substrate in the presence of a fixed concentration of inhibitor. A reciprocal plot of the first will provide the value of K_m of the substrate, and a reciprocal plot of the second will provide the value of K_a (Apparent K_m) in the presence of the particular concentration of inhibitor used. Since the latter constant is equivalent to $K_m(1+([i]/K_i))$, this relationship can be expressed by an equation:

$$K_a = K_m \left(1+\frac{[i]}{K_i}\right) \tag{87}$$

and the value of K_i found by inserting the values of the remaining three terms.

A graph of this type of analysis is shown in Fig. 34, which is a reciprocal plot of Curves A and B in Fig. 30. The two lines representing transport in the presence and absence of inhibitor cross the ordinate at the same point, which represents $1/v_{max}$. The points on the abscissa at a value of $1/\frac{1}{2}.v_{max}$ on the ordinate correspond to $1/K_a$ and $1/K_m$, $1/K_a$ having the smaller value (K_a K_m).

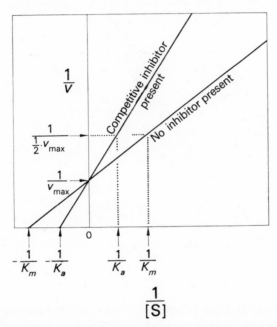

Fig. 34. Double reciprocal plot showing the effect of a fixed concentration of competitive inhibitor on the rate of transfer (v) of substrate at various concentrations ([S]) (cf. Figs. 20 and 23). The value for $1/v_{max}$ is unchanged by the presence of the inhibitor, but that for $1/K_m$ appears reduced (i.e. K_m appears increased) to $1/K_a$, where

$$K_a = K_m \left(1+\frac{[i]}{K_i}\right).$$

Competitive inhibition can be identified by intersection of the lines on the ordinate, indicating an unchanged value for v_{max}, as shown in Fig. 34. In non-competitive inhibition they intersect at the abscissa, indicating an unchanged value for K_m. This is shown in Fig. 35, where similar triangles have been shaded to demonstrate the geometry. Fig. 36 shows an experimental example of the use of this technique.

72 CHAPTER 5

Plot of v against $\dfrac{v}{[S]}$ [50]

The procedure of plotting v against $v/[S]$ (p. 43) can be used to determine kinetic constants in the presence of a fixed concentration of inhibitor. The rationale is identical with that on which the use of the double-reciprocal plot (p. 70) is based. Two straight-line plots are compared, one representing transfer of substrate in the presence of inhibitor, the other showing transfer in its absence.

With a competitive inhibitor the value of v_{max} is unchanged, but that of K_m is increased; the two lines obtained will thus meet on the ordinate at a single point which represents v_{max} (Fig. 37). The slopes are not the same since they depend on the difference in the values attributed to K_m. In Fig. 37, $K_m = (C/D)$ and $K_a = (C/E)$. The lower

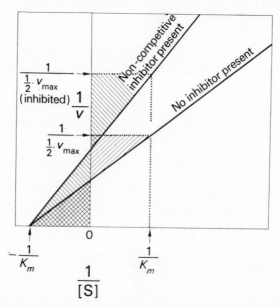

Fig. 35. Double reciprocal plot showing the effect of a fixed concentration of non-competitive inhibitor on the rate of transfer (v) of substrate present at various concentrations ([S]) (cf. Figs. 20 and 23). The value of $1/v_{max}$ is increased (i.e. v_{max} is decreased) as shown by an increase in the value ($1/v$) at which the plotted line crosses the ordinate, but the value of $1/K_m$ is unchanged. The *shaded areas* represent two sets of similar triangles to show that both plotted lines cut the abscissa at $-(1/K_m)$.

value of E as compared with D results in a relatively higher value for K_a as compared with the true K_m. K_i can then be determined as in the double-reciprocal plot by inserting the known values into equation (87).

If inhibition is non-competitive, as shown in Fig. 38, the line representing inhibited transport will be parallel to that representing uninhibited transport, since the value for K_m will be the same in each case. Intersection with the ordinate will be at a lower value, representing a lower value for v_{max}.

VARIATION OF CONCENTRATION OF INHIBITOR WITH
FIXED CONCENTRATION OF SUBSTRATE

K_i can also be determined by using two fixed concentrations of substrate and varying the concentration of inhibitor with each.

Plot of rate of transport against inhibitor concentration

In this method, the concentration of substrate is fixed, and that concentration of inhibitor is found which reduces the rate of transport of substrate by half. This is repeated using a second fixed concentration of substrate, and K_i can then be found by calculation as shown below.

When substrate is present at a concentration, $[S]_1$, without inhibitor it is transported at a rate v_1. This rate can also be expressed as

$$\frac{v_{max} \cdot [S]_1}{[S]_1 + K_m}$$

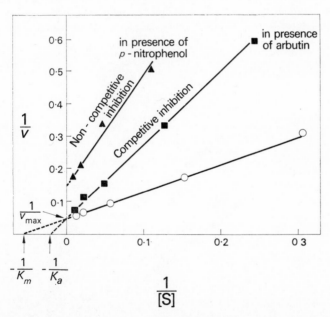

Fig. 36. Double reciprocal plot of experimental data showing competitive and non-competitive inhibition. Effect of 2·5 mM arbutin (competitive inhibitor) and of 2·5 mM p-nitrophenol (non-competitive inhibitor) on uptake of 6-deoxyglucose by intestine *in vitro*. (Arbutin is a derivative of glucose, and hence is also structurally related to 6-deoxyglucose, thus explaining the competitive nature of its inhibition.) Compare with Figs. 34 and 35. *Open circles* represent uptake of 6-deoxyglucose in the absence of inhibitor. v = rate of uptake of 6-deoxyglucose, μmoles/ml tissue water/10 min. $[S]$ = external concentration of 6-deoxyglucose, mM. Incubation 10 min, 37°C. $K_m = 2$ mM, K_a (eqn. 87, p. 71) of 6-deoxyglucose in the presence of arbutin = 5 mM. (Redrawn from data published by Alvarado & Crane [3].)

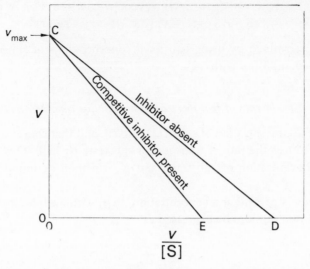

Fig. 37. Plot of v against $v/[S]$ showing the effect of a fixed concentration of competitive inhibitor on the rate of transfer (v) of substrate at various concentrations ([S]) (cf Figs. 21 and 24). The value of v_{max} is unchanged but the ratio $v : (v/[S])$, i.e. the slope of the plotted line, is altered from C/D to C/E, indicating an apparent change in the value of K_m.

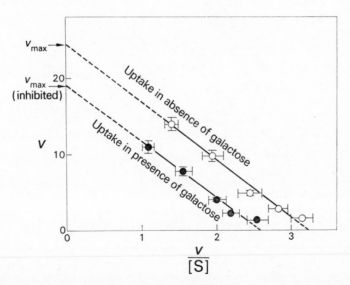

Fig. 38. Plot of v against $v/[S]$ from experimental data showing non-competitive inhibition (cf. Figs. 21 and 24). Uptake of valine by intestine *in vitro* in the presence and absence of 25 mM galactose. v = intracellular concentration of valine (mM) after 5 min. [S] = concentration of valine in suspending medium, mM. Each point represents the mean of six values for v ($\pm$s.e.m.) at each value of [S]. Incubation 5 min, 37°C. The slopes of the plotted lines are similar, showing an unchanged value for K_m; the value of v_{max} was reduced. (Drawn from values published by Reiser & Christiansen [124].)

(eqn. 27). Using the same concentration of substrate, values for the transport rate (v) are determined in the presence of several concentrations of competitive inhibitor and plotted against the concentration of inhibitor ([i]) as shown in Fig. 39. (In this example, the concentration of substrate used ([S]$_1$) is equal to K_m.) From the graph a concentration of inhibitor is then found ([i]$_1$) which reduces the rate of transport of substrate to $\frac{1}{2} . v_1$. This rate, of course, is equal to

$$\frac{1}{2} . \frac{v_{max} . [S]_1}{[S]_1 + K_m},$$

and is also equal to

$$\frac{v_{max} . [S]_1}{[S]_1 + K_m \left(1 + \dfrac{[i]_1}{K_i}\right)}$$

since this represents the rate of transport at that particular concentration of substrate and that particular concentration of inhibitor (eqn. 69, p. 60).

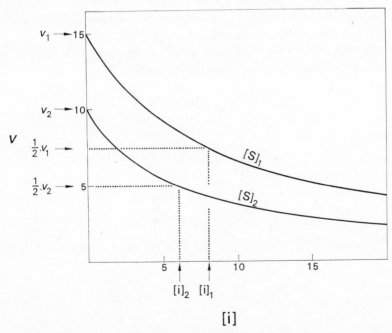

Fig. 39. The determination of K_m and K_i by finding the concentration of competitive inhibitor required to reduce the rate of carrier transport of substrate by half. Rate of transport of substrate at two concentrations ([S]$_1$ and [S]$_2$) is determined in the absence of inhibitor, and then the concentration of inhibitor found ([i]$_1$ and [i]$_2$ respectively) which reduces each rate to half its uninhibited rate. From these values K_m and K_i may be determined as described in the text. Units are arbitrary: $K_m = 1$, $K_i = 4$, $v_{max} = 30$, [S]$_1$ ($= K_m$) $= 1$, [S]$_2$ ($= (K_m/2)$) $= 0.5$. v = rate of transfer of substrate; [i] = concentration of inhibitor.

Hence

$$\tfrac{1}{2} \cdot \frac{v_{max} \cdot [S]_1}{[S]_1 + K_m} = \frac{v_{max} \cdot [S]_1}{[S]_1 + K_m \left(1 + \dfrac{[i]_1}{K_i}\right)} \tag{88}$$

which simplifies to

$$2([S]_1 + K_m) = ([S]_1 + K_m) + \frac{K_m \cdot [i]_1}{K_i} \tag{89}$$

and further to

$$[S]_1 + K_m = \frac{K_m \cdot [i]_1}{K_i} \tag{90}$$

from which values for K_m and K_i can be obtained thus:

$$K_m = \frac{[S]_1 \cdot K_i}{[i]_1 - K_i} \tag{91}$$

$$K_i = \frac{K_m \cdot [i]_1}{[S]_1 + K_m}. \tag{92}$$

The same procedure is repeated, but at a different concentration of substrate, $[S]_2$, which results in a rate of transport, v_2, in the absence of inhibitor. The concentration of inhibitor found to reduce this rate by half is $[i]_2$ (Fig. 39).

There will now be two sets of experimental values, and two equations for K_m, in which the values of [S] and [i] in both cases are known; i.e.

$$K_m = \frac{[S]_1 \cdot K_i}{[i]_1 - K_i} \text{ and } K_m = \frac{[S]_2 \cdot K_i}{[i]_2 - K_i} \tag{93}$$

so that

$$\frac{[S]_1 \cdot K_i}{[i]_1 - K_i} = \frac{[S]_2 \cdot K_i}{[i]_2 - K_i} \tag{94}$$

hence

$$K_i = \frac{[S]_1 \cdot [i]_2 - [S]_2 \cdot [i]_1}{[S]_1 - [S]_2}. \tag{95}$$

Similarly, K_m can be found, for there are also two separate equations describing K_i, thus:

$$K_i = \frac{K_m \cdot [i]_1}{[S]_1 + K_m} \text{ and } K_i = \frac{K_m \cdot [i]_2}{[S]_2 + K_m} \tag{96}$$

whence

$$K_m = \frac{[S]_1 \cdot [i]_2 - [S]_2 \cdot [i]_1}{[i]_1 - [i]_2} \tag{97}$$

The value of v_{max} may now be found by inserting the known values of [S], v and K_m into the basic kinetic equation (eqn. 27).

In these equations there is the risk of cumulative error, which could be particularly great in the determination of K_m and hence of v_{max}. If it is assumed that the error in [S] is small enough to be ignored, there are two variables, $[i]_1$ and $[i]_2$. In equation (95), which describes K_i, these are confined to the numerator, so that cumulative error is unlikely to be large. In equation (97), however, they are also present in the denominator, so that the error in the value obtained for K_m could be very large indeed.

This method for finding K_i can be simplified if the concentration of substrate is made *very small* compared with the value of K_m. Then the term ([S]+K_m) simplifies to K_m without serious error. Only one set of values need then be obtained since equation (90) becomes

$$[i]_1 = K_i. \tag{98}$$

In other words, when [S] is very small compared with K_m, K_i is that concentration of inhibitor which reduces the rate of transport of substrate by half.

The method described here is based on the assumption that inhibition is competitive; it cannot distinguish competitive from non-competitive inhibition. If inhibition is non-competitive, $v_{\max}$ is reduced and the inhibition equation (eqn. 69) does not apply, so that the equation (eqn. 88) upon which the method is based cannot be used.

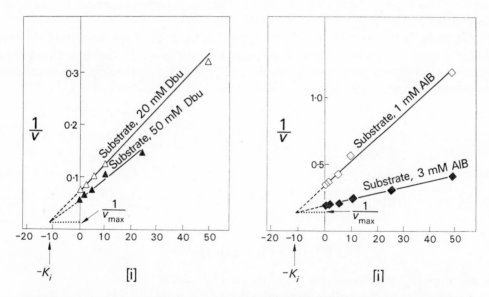

Fig. 40. Determination of K_i from a plot of $1/v$ against [i]. The amino acid phenylalanine inhibited competitively the transfer of α,γ-diaminobutyrate (Dbu) (*left*) and of α-amino-isobutyrate (AIB) (*right*) into Ehrlich ascites carcinoma cells. In each case, K_i for phenylalanine was found to be about 11 mM, which is consistent with a single common carrier for phenylalanine, Dbu and AIB. (However, when phenylalanine acts as inhibitor of the transfer of the amino acids L-leucine and L-valine in the same tissue, its K_i has then been found to be about 0·5 mM (Table 4), suggesting that under those circumstances more than one carrier system was involved (see Chapter 6, p. 80). v=mmoles of substrate/kg cell water/min. [i] = concentration of phenylalanine in suspending medium, mM. In each case $v_{\max}$ refers to corresponding substrate. Incubation at 37°C: α,γ-diamino-butyrate, 2 min; α-aminoisobutyrate, 1 min. (Redrawn from data published by Christensen & Liang [35].)

Plot of $1/v$ against [i] [49, 50]

The experimental part of the technique in this method is similar to that in the immediately preceding method, but here $1/v$ is plotted against [i]. If Michaelis–Menten kinetics apply, the plotted line will be straight, as can be shown by inverting the inhibition equation (69) thus:

$$\frac{1}{v} = \frac{[S] + K_m + \dfrac{K_m.[i].}{K_i}}{v_{\max}.[S]}. \tag{99}$$

This can be broken into its component parts:

$$\frac{1}{v} = \frac{[S] + K_m}{v_{\max}.[S]} + \frac{K_m.[i]}{K_i.v_{\max}.[S]}. \tag{100}$$

Since all values except v and [i] are constant, $1/v$ is proportional to [i], indicating a linear function.

If the concentration of inhibitor is varied with each of two concentrations of substrate, and $1/v$ plotted against [i] in each case, two straight lines will be obtained. The value on the abscissa where these cross is equal to K_i (Fig. 40).

Algebraically it may be put as follows [50]. At the point of intersection of the two lines, the value of v and of [i] will be the same for both lines. The inhibition equations (eqn. 69) applicable to each can be combined, $[S]_1$ and $[S]_2$ representing the two concentrations of substrate used, thus:

$$\frac{v_{\max}.[S]_1}{[S]_1 + K_m\left(1 + \dfrac{[i]}{K_i}\right)} = \frac{v_{\max}.[S]_2}{[S]_2 + K_m\left(1 + \dfrac{[i]}{K_i}\right)} \tag{101}$$

hence

$$[S]_1.[S]_2 + [S]_1.K_m + \frac{[S]_1.K_m.[i]}{K_i} = [S]_1.[S]_2 + [S]_2.K_m + \frac{[S]_2.K_m.[i]}{K_i} \tag{102}$$

This simplifies to

$$[S]_1\,(K_i + [i]) = [S]_2\,(K_i + [i])$$

which, since $[S]_1$ is different from $[S]_2$, can only be valid if $[i] = -K_i$.

$v_{\max}$ can also be determined since the value of $1/v$ at the point of intersection of the lines can be shown to represent $1/v_{\max}$. At this point, $[i] = -K_i$, and if this is inserted into the general inhibition equation (eqn. 69, p. 60), the latter becomes:

$$v = \frac{v_{\max}.[S]}{[S] + K_m\left(1 - \dfrac{K_i}{K_i}\right)}. \tag{103}$$

The reciprocal of this after simplification is then

$$\frac{1}{v} = \frac{1}{v_{\max}}. \tag{104}$$

If all known values are now inserted into the equation (eqn. 69), K_m is obtained.

This method is also used in enzyme kinetics to identify the effect of a non-competitive inhibitor; the two lines then meet on the abscissa [50]. However, the method does not distinguish between pure competitive inhibition and a mixture of the two types, unless v_{max} has first been determined with a reliability which is not as common as might be supposed. Only if the lines cross at a value of $1/v$ equal to $1/v_{max}$ can it be assumed that inhibition is solely competitive.

Figure 40, which is an example of this kind of analysis, is of particular interest since it also demonstrates constancy of K_i for a particular inhibitor, as would be expected if it was identical to its K_m. The figure shows the amino acid phenylalanine as a competitive inhibitor of two other amino acids, diaminobutyrate and α-aminoisobutyrate. The value for the K_i of phenylalanine is the same with either substrate, having a value of about 11 mM. (Its K_i as an inhibitor of leucine and valine, however, is about 0·5 mM, a difference which may be explained in terms of more than one carrier system (see Chapter 6).)

Algebraical method

K_i may be determined by means of a 'velocity ratio' [50] relating the uninhibited rate of transfer to the inhibited rate, provided that inhibition is competitive. The rate is determined in each case using the same concentration of substrate, and the values inserted into the following equation, which is obtained by dividing equation (27) by equation (69) and cancelling out $v_{max}.[S]$:

$$\frac{v}{v_{(i)}} = \frac{[S]+K_m\left(1+\frac{[i]}{K_i}\right)}{[S]+K_m} \tag{105}$$

where v and $v_{(i)}$ are the rates of transfer in the absence and presence respectively of competitive inhibitor. The equation can be rearranged to the following form:

$$K_i = \frac{v_{(i)}}{v-v_{(i)}} \cdot \frac{K_m.[i]}{[S]+K_m}. \tag{106}$$

K_m must be found independently and so its reliability will affect the accuracy of the value obtained for K_i.

Chapter 6

Multiple Carriers

Sometimes the measurement of 'initial rate' as an indication of one-way transfer does not give experimental data which fit the single carrier hypothesis [18, 36, 37, 52, 107, 114, 132]. It may be reasonable then to consider that more than one carrier might be present in such cases, and to ask whether an appropriate adaptation of Michaelis–Menten kinetics would explain the observed results. Frequently, there is a satisfactory correspondence between observation and what may be expected from the adapted kinetics. It is therefore worth looking at the rationale in these circumstances, and discussing the simplest case, in which it is assumed that two carriers, with different kinetic properties, both transfer a single substrate simultaneously.

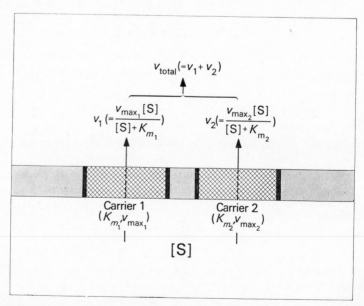

Fig. 41. Two-carrier model. Total rate of one-way carrier transport (v_{total}) is the sum of the rates of transport of each carrier separately ($v_1 + v_2$): each of these is determined by the concentration of substrate ([S]) and the kinetic constants associated with the corresponding carrier (K_{m_1} and v_{max_1}; K_{m_2} and v_{max_2}).

Kinetics of a two-carrier system

If we assume that there are two carriers acting in parallel, the total one-way rate of transfer (ignoring diffusion) may be expected to be the arithmetical sum of the two component rates, as shown in Fig. 41, since the quantity transferred by two carriers in unit time must be the sum of the masses transported by each. Algebraically this appears as:

$$v_{\text{total}} = v_1 + v_2 = \frac{v_{\text{max}_1} \cdot [S]}{[S] + K_{m_1}} + \frac{v_{\text{max}_2} \cdot [S]}{[S] + K_{m_2}}, \tag{107}$$
$$\text{(Carrier 1)} \quad \text{(Carrier 2)}$$

where the subscripts '$_1$' and '$_2$' refer to the respective carriers, and the whole is the simple sum of two orthodox carrier equations.

A relationship of this sort can be seen in Fig. 42 which shows experimental data from what is believed to be a dual carrier system of this sort. The data are derived from an investigation into the uptake of potassium by barley roots [54], the circles representing the experimental values.* In addition, curves representing two hypothetical single carriers (interrupted lines) have been drawn, one corresponding to a carrier with a K_m of 0·02 mM and a v_{max} of 11 μmoles/g/h (Carrier 1) and the other to a carrier with a K_m of 10 mM and a v_{max} of 14 μmoles/g/h (Carrier 2). If the values for v on the two curves at each value of [S] are added together, the continuous line is obtained. This corresponds remarkably well to the experimental values, which suggests that potassium was taken up by a two-carrier system. In support of this, no adequate fit of the data to a curve derived from a single carrier assumption was found possible.

A composite curve, like that in Fig. 42, derived from the equation (eqn. 107) for a two-carrier system, does not show the symmetry of the curve associated with a one-carrier system; when multiplied out, the equation (eqn. 107) is revealed as a quadratic function. Although it has been suggested elsewhere that a curve describing a multiple-carrier system of this sort should show apparent discontinuities, there is no point at which a sudden alteration of the rate of change of slope would be expected. The rate of rise of both types of curve (representing one and two-carrier systems) shows a continuously progressive reduction with rise in concentration.

Break-down of the data from a two-carrier system into two components is usually not difficult, provided that the K_m values are widely separated as in this case. Preliminary kinetic constants for one of the carriers are first determined on the assumption that a portion of the curve for total transport approximates to the contribution by one of the

---

* Other experiments suggest that the entry of potassium into barley roots may in certain conditions be even more complex than suggested here [52]. The data of Fig. 36 come from experiments in which the only solutes added to the water bathing the roots were KCl and CaCl₂ (CaCl₂, 0·5 mM). When CaSO₄ (0·5 mM) replaced the CaCl₂ a step-wise increase in transfer rate appeared to be superimposed upon that shown here as the concentration of potassium was raised, with the rate at 50 mM KCl being approximately doubled [53]. However, a close examination of published results [53] shows that there are insufficient experimental values in relation to experimental error to allow a firm conclusion as to the reality of a step-wise appearance.

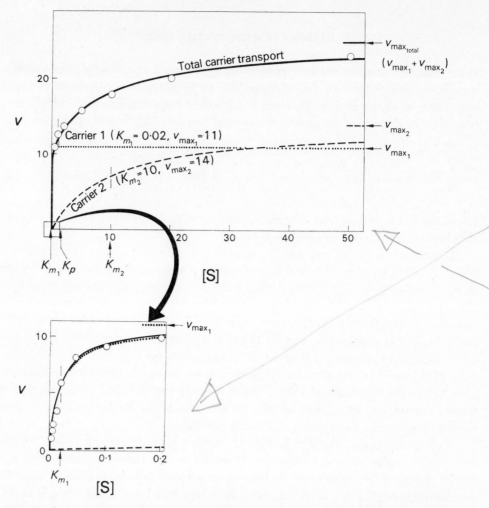

Fig. 42. Two-carrier transport. The uptake of labelled potassium by barley roots interpreted as a two-carrier system, one carrier (Carrier 1) having a K_m of 0·02 mM and a
v_{max} of 11 μmoles/g fresh weight of tissue/h, the other (Carrier 2) a K_m of 10 mM and a
v_{max} of 14 μmoles/g fresh weight of tissue/h. The kinetic curve for each carrier is shown
as an *interrupted line*; the *solid line* represents the values obtained by adding the separate
components together, and closely follows the experimental values (shown by *circles*).
Curves calculated from equation (107). The *expanded portion* shows the lower concentration range in more detail; at these concentrations, total uptake approximates to uptake
by Carrier 1. Potassium uptake measured as ^{42}K from a solution of ^{42}KCl in 0·5 mM
CaCl$_2$. v = rate of uptake of potassium (μmoles/g fresh weight of tissue/h). [S] = external concentration of potassium (mM). K_p = concentration of potassium which produced
half the estimated maximal rate of total transfer. Incubation 10 min, 30°C. (Published
constants: K_{m_1} = 0·021 mM, v_{max_1} = 11·9 μmoles/g fresh weight of tissue/h; K_{m_2} =
11·4 mM, v_{max_2} = 13·2 μmoles/g fresh weight of tissue/h [54].) (Drawn from data
published by Epstein *et al.* [54].)

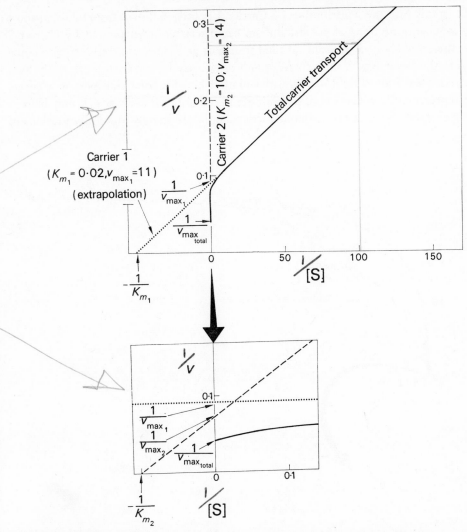

Fig. 43. Double reciprocal plot of the curves shown in Fig. 42. The *main chart* corresponds to the *expanded portion* of Fig. 42, which covers the lower concentration range; the range $(1/[S]) = 10-150$ mM^{-1} is roughly equivalent to $[S] = 0.1-0.007$ mM and the range $(1/v) = 0.1-0.3$ (μmoles/g fresh weight/h)$^{-1}$ is roughly equivalent to $v = 10-3$ μmoles/g fresh weight/h. The straight part of the plotted line for total transport can be extrapolated to provide values for $1/v_{\text{max}_1}$ and $(-1/K_{m_1})$ since in the corresponding concentration range total carrier transport approximates to transport by Carrier 1. The *expanded portion* of this figure covers the higher concentration range and corresponds to the *main chart* of Fig. 42: the range $(1/[S]) = 0.04-0.1$ mM^{-1} is equivalent to $[S] = 25-10$ mM and the range $(1/v) = 0.04-0.1$ (μmoles/g fresh weight/h)$^{-1}$ to $v = 25-10$ μmoles/g fresh weight/h. This region cannot be used for obtaining any kinetic constants directly, since at no point does the line representing total transport approximate to that representing transport by either of the individual carriers. The reason should be obvious from an examination of Fig. 42. Symbols as in legend to Fig. 42.

carriers. In this case the contribution by Carrier 1 is little different from total transport at very low concentrations of substrate (Fig. 42, inset), the contribution by Carrier 2 representing only about 5 per cent. of total transport at a substrate concentration equal in value to the lower K_m. Values for K_{m_1} and v_{max_1} may be obtained more easily by a double reciprocal plot (Fig. 43). Here, the plotted line for total transport is a curve, since it represents the reciprocal of a quadratic function (a property which may help in detecting the presence of more than one carrier). At appropriately low concentrations of

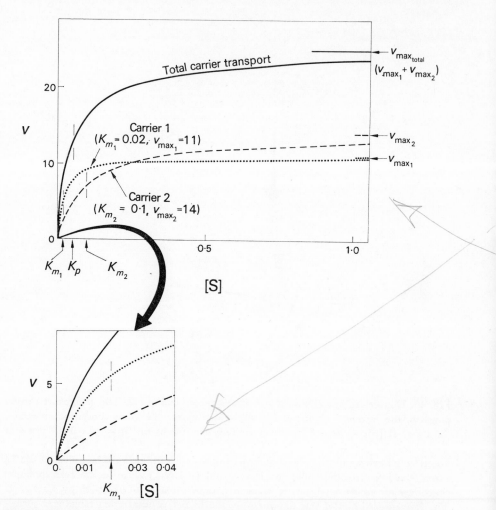

Fig. 44. Transport by a hypothetical two-carrier system having the same parameters as those in Fig 42 except that K_m for Carrier 2 is 0·1 mM, that is, one-hundredth of that in Fig. 42 and only five times the K_m for Carrier 1. The *expanded portion* shows details of transport at low concentrations. At no point does the curve for total transport approximate to that of either of the carriers; it would therefore not be possible for the kinetic constants of either carrier to be obtained by direct analysis of the plot. The concentration (K_p) which produces half the maximal rate of total transport lies between the value of K_{m_1} and K_{m_2}. Symbols as in legend to Fig. 42. Curves calculated from equation (107).

substrate (high values of 1/[S]) the reciprocal plot is approximately straight and, if extrapolated, may provide rough values for the kinetic constants of Carrier 1, since at these concentrations the values for total transport may approximate to those of Carrier 1. When these constants have been obtained, approximate transport rates for Carrier 1

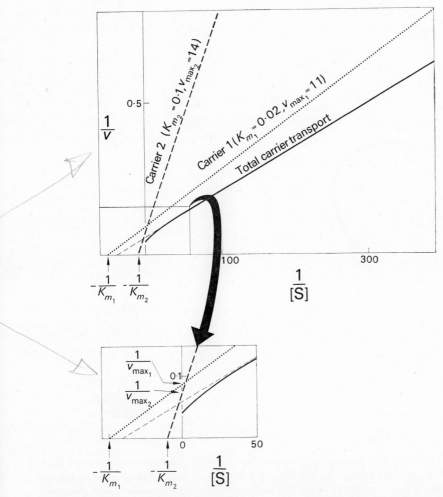

Fig. 45. Double reciprocal plot of the curves shown in Fig. 44. The *main chart* corresponds to the *expanded portion* of Fig. 44, which covers the lower concentration range; the range (1/[S]) = 30–300 mM^{-1} is roughly equivalent to [S] = 0·03–0·003 mM and the range (1/v) = 0·04–0·5 (μmoles/g fresh weight/h)$^{-1}$ to v = 25–2 μmoles/g fresh weight/h. The *expanded portion* of this figure covers the higher concentration range and corresponds to the *main chart* of Fig. 44: the range of (1/[S]) = 0–50 mM^{-1} is equivalent to [S] = ∞– 0·02 mM and the range of (1/v) = 0·04–0·1 (μmoles/g fresh weight/h)$^{-1}$ to v = 25– 10 μmoles/g fresh weight/h. As would be expected from an examination of Fig. 44, at no point does the line representing total transport approximate to either of the lines representing the individual carriers, and hence no kinetic constants can be obtained directly. Symbols as in legend to Fig. 42.

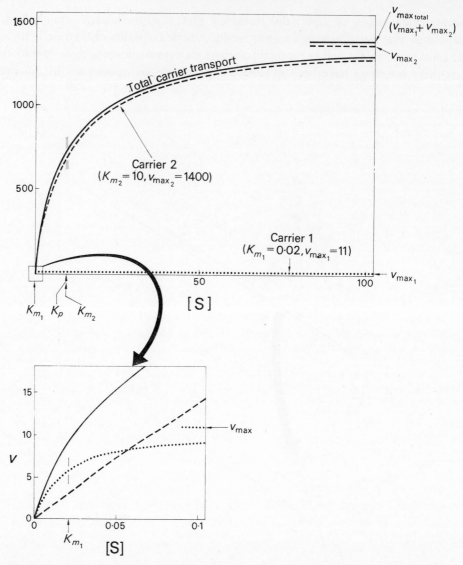

Fig. 46. Transport by a hypothetical two-carrier system having the same parameters as those in Fig. 42 except that v_{max} for Carrier 2 is a hundred times greater, i.e. v_{max_2} = 1400 μmoles/g fresh weight of tissue/h. Over the higher concentration range (between [S] = 1 mM and [S] = 100 mM) total transfer would approximate to transfer by Carrier 2, so that its analysis would provide values for kinetic constants not very different from those of Carrier 2, and the value of K_p would be almost identical with that of K_{m_2}. The *expanded portion* shows details of transport at concentrations of substrate below 0·1 mM; at these concentrations the behaviour of Carrier 2 is indistinguishable from diffusion and may be treated as such, so that a measure of the constants for Carrier 1 might be obtained after subtracting an apparently unsaturable component. Symbols as in legend to Fig. 42. Curves calculated from equation (107).

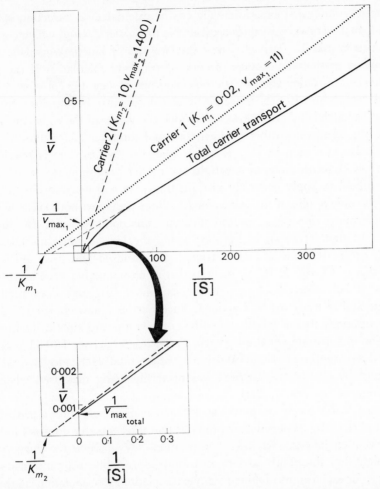

Fig. 47. Double reciprocal plot of the curves shown in Fig. 46. The *main chart* corresponds to the *expanded portion* of Fig. 46, which covers the lower concentration range; the range $(1/[S]) = 10–300$ mM^{-1} is roughly equivalent to $[S] = 0·03–0·003$ mM and the range $(1/v) = 0·07–0·5$ (μmoles/g fresh weight/h)$^{-1}$ to $v = 15–2$ μmoles/g fresh weight/h. As would be expected from an examination of Fig. 46, nowhere is there a similarity between either of the lines representing the two carriers and the line representing total transport in this concentration range. The *expanded portion* of this figure covers the higher concentration range and corresponds to the *main chart* of Fig. 46: the range $(1/[S]) = 0·01–0·3$ mM^{-1} is roughly equivalent to $[S] = 100–3$ mM and the range $(1/v) = 0·0007–0·002$ (μmoles/g fresh weight/h)$^{-1}$ to $v = 1500–500$ μmoles/g fresh weight/h. In this concentration range the line representing total transport approximates to the line representing Carrier 2, but only in the range $(1/[S]) = 0–0·1$ mM^{-1} ($[S] = \infty–10$ mM), i.e. at concentrations above the value of K_{m_2}; it might in theory be possible to obtain a value for $-(1/K_{m_2})$ and of $1/v_{\max_2}$ by extrapolation. (The line representing Carrier 1 is not shown in the expanded diagram; it would cross the ordinate at $(1/v) = 0·091$ (μmoles/g fresh weight/h)$^{-1}$.) Symbols as in legend to Fig. 42.

at the concentrations used experimentally can be calculated. Subtraction of these from the experimental values gives values attributable to Carrier 2, from which corresponding kinetic constants may be derived. From the two sets of kinetic constants, two sets of transfer rates over the whole range of concentration are calculated. If the sum of the rates at each concentration agrees with the experimental data, the constants used may be considered acceptable. If not, more accurate values must be obtained by successive approximation, a trial and error process in which the values of the constants are adjusted until there is good agreement between experimental and calculated values.

In the example shown in Fig. 42, analysis of the two carriers is relatively simple because at low concentrations of substrate one carrier predominates, a situation which is often assumed to apply generally in both transport and enzyme kinetics [50, 137]. However, this only applies if the values for K_m differ widely without excessive difference between the values for v_{max}, and to illustrate the point we give two hypothetical examples. The first is shown in Fig. 44, where the kinetic constants used are those of Fig. 42, except that the K_m of Carrier 2 is a hundred-fold less, and is now only five times that of Carrier 1. It can be seen that at no point (not even at very low concentrations of substrate, as shown in the expanded diagram) does total transport approximate well to transport by Carrier 1, since Carrier 2 now contributes appreciably at all concentrations. It can be calculated from the composite kinetic equation that at a concentration of substrate equal once again to the K_m of Carrier 1 (0·02 mM), transport by Carrier 2 now represents about 30 per cent. of the total. Hence to have assumed here that transport by one of two carriers was primarily being measured when substrate concentration was low would clearly be misleading.

Nor does a double reciprocal plot (Fig. 45) help analysis here since it provides nothing new except to show by its curvilinear property that total transfer does not follow simple Michaelis–Menten kinetics. At low concentrations of substrate the curve is approximately straight, but since both carriers contribute appreciably to total transport, neither K_m nor v_{max} for each can be obtained from extrapolation. Extrapolation (to the ordinate) of data obtained at very high concentrations of substrate (Fig. 45, *expanded portion*) may, however, supply a value for the maximal rate of transfer by the system as a whole.

The second hypothetical example is shown in Fig. 46, in which the values for the constants are also those of Fig. 42 except that v_{max_2} is very large, with a value a hundred-fold greater than that in Fig. 42, that is, about 130 times greater than v_{max_1}. It can be seen in Fig. 46 that at high concentrations of substrate the rate of transport by Carrier 2 approximates to the rate of total transport and that that by Carrier 1 is relatively very small. Hence over the higher range of concentrations approximate values for the constants of Carrier 2 may be obtained from the values for total transfer with the help, perhaps, of a double reciprocal plot (Fig. 47, *expanded portion*), but there is here no concentration of substrate at which the constants for Carrier 1 may be obtained directly (compare expanded portions of Figs. 42 and 46). It may be difficult to separate the component with the lower v_{max} (Carrier 1), since in determining the characteristics of Carrier 2 the contribution by Carrier 1 would have been treated as if it were zero at all concentrations. However, at very low concentrations of substrate, transport by Carrier 2 cannot be distinguished from diffusion; if, therefore, at these concentrations, an apparently unsaturable component could be subtracted from total transport, values for Carrier 1 might be obtained.

If the two values for K_m in a two-carrier system are the same, the two carrier components behave together as if a single carrier only were present, whether or not v_{max_1} and v_{max_2} are equal. They are then indistinguishable from a single carrier by orthodox analysis.

In a two-carrier system, the concentration at which the total rate of transport is half-maximal should not, strictly speaking, be identified with the term K_m (a constant applicable only to simple Michaelis–Menten kinetics); although this concentration represents a constant for the system as a whole, it is not a constant for individual participants. We therefore prefer to use the term 'Pseudo-K_m' or K_p, to describe this value. It is of course related, but in a complex way, to the kinetic constants of each carrier. This relationship may be found as follows.

The concentration of substrate which produces half the total maximal rate of transport i.e.

$$\left(\frac{v_{max_1} + v_{max_2}}{2}\right)$$

is referred to as K_p. Let us assume that each carrier is exposed to this concentration; then in the two-carrier equation (eqn. 107) we can say that when

$$v_{total} = \frac{v_{max_1} + v_{max_2}}{2}, \quad [S] = K_p.$$

K_p may thus be substituted for [S] in this equation (eqn. 107) which then appears as:

$$\frac{v_{max_1} + v_{max_2}}{2} = \frac{v_{max_1}.K_p}{K_p + K_{m_1}} + \frac{v_{max_2}.K_p}{K_p + K_{m_2}}. \tag{108}$$

Hence,

$$v_{max_1}(K_p + K_{m_1})(K_p + K_{m_2}) + v_{max_2}(K_p + K_{m_1})(K_p + K_{m_2}) \tag{109}$$

$$= 2v_{max_1}.K_p(K_p + K_{m_2}) + 2v_{max_2}.K_p(K_p + K_{m_1})$$

and

$$2v_{max_1}.K_p(K_p + K_{m_2}) - v_{max_1}(K_p + K_{m_1})(K_p + K_{m_2}) \tag{110}$$

$$= v_{max_2}(K_p + K_{m_1})(K_p + K_{m_2}) - 2v_{max_2}.K_p(K_p + K_{m_1})$$

so that

$$\frac{v_{max_1}}{v_{max_2}} = \frac{(K_p + K_{m_1})(K_p + K_{m_2}) - 2K_p(K_p + K_{m_1})}{2K_p(K_p + K_{m_2}) + (K_p + K_{m_1})(K_p + K_{m_2})} \tag{111}$$

i.e.

$$\frac{v_{max_1}}{v_{max_2}} = \frac{(K_p + K_{m_1})(K_{m2} - K_p)}{(K_p + K_{m_2})(K_p - K_{m_1})} \tag{112}$$

It is clear from this that K_p cannot be smaller than K_{m_1} (for v_{max_1}/v_{max_2} is a positive number); also K_p cannot be greater than K_{m_2} for the same reason.

If $v_{max_1} = v_{max_2}$, equation (112) states that $K_p = \sqrt{(K_{m_1}.K_{m_2})}$ (geometric mean).

(It also follows that under these circumstances, if K_{m_1} were equal to K_{m_2},

$$K_p = K_{m_1} = K_{m_2},$$

which effectively represents a single carrier.)

The value of K_p depends upon the particular circumstances. For example, in Fig. 42 it has a value of 1·5 mM, as compared with 0·02 and 10 mM for the K_m values for the two carriers; in Fig. 44 it is about half-way between the two K_m values, and in Fig. 46 it is close to the K_m of the carrier with the higher v_{max}.

The double reciprocal plot, a method of analysis often used in attempting to extract the kinetic parameters of two-carrier systems [132, 137], must be used with discretion, as suggested by Figs. 42–47. In some published work, straight lines have been drawn through the plotted points at each end of a curved double reciprocal plot [137, 163], which may give the misleading impression that each line describes the kinetics of a single carrier, and values termed v_{max} and K_m have in fact been derived from them directly [141]. Although it may be reasonable in certain circumstances to use the upper of two such lines in this way (as in Fig. 43), the lower one usually has no quantitative meaning.

A double reciprocal plot can also be used as an aid in determining whether more than one carrier system is present, since the plot will then be curved, in contrast to the straight-line plot of a single carrier conforming to Michaelis–Menten kinetics. However, before analysing a curve of this sort, it is important that any component of diffusion is either negligible, or has first been subtracted from the data, since diffusion will constitute a third component of transfer. Furthermore, a one-carrier system in the presence of diffusion also leads to a curved double reciprocal plot, which is approximately linear at low concentrations of substrate (cf. Figs. 23 and 47) and so may be confused with that derived from a two-carrier system. However, when diffusion is present, the plotted line passes through the origin, but when absent the line cuts the ordinate above the origin. The plotted line may also *appear* to pass through the origin in the absence of diffusion if the concentration range over which transfer is analysed is limited, as shown in Fig. 47.

In summary, it is not difficult to analyse the behaviour of a model system containing two carriers, but in practice reliable conclusions can only be reached if the sum of the numerical values for each carrier which have been derived from the kinetic analysis give a good fit to the experimental values. The analysis usually entails a series of successive approximations, but only where the values for K_m are widely separated may the derivations be relatively simple and the values reasonably reliable.

Kinetics of inhibition

The principles of inhibition of multiple carriers in parallel are identical with those for single carriers, but analysis is more complex since the number of variables is larger. As before, we shall deal with the simplest case which involves only two carriers.

An inhibitor may act on one of two carriers only, or on both. If it acts on only one, the problem is relatively simple. By increasing the concentration of inhibitor to levels at which the inhibition of transport of substrate by one carrier is virtually complete, the behaviour of the other carrier may be determined, its contribution to total transfer may then be predicted for various experimental conditions, and the contribution by the first carrier derived for these by subtraction. An example of this, using a competitive inhibitor, is shown in Fig. 48. Here the effect of increasing concentrations of the amino acid α-(methylamino)isobutyrate on the uptake of histidine into Ehrlich ascites carcinoma cells is illustrated. A 10 mM concentration of the inhibitor reduced the uptake of histidine by about one-third. Increasing the concentration to 50 mM had no obvious further effect, suggesting that one component of transfer had been virtually saturated by

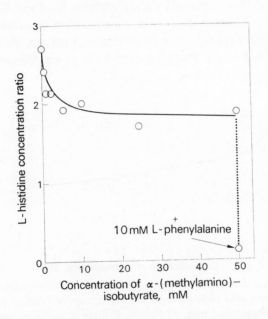

Fig. 48. Evidence for a two-carrier transfer system. Apparent saturation of part of the uptake of the amino acid L-histidine into Ehrlich ascites carcinoma cells by α-(methyl-amino)isobutyrate at high extracellular concentrations, and inhibition of the unaffected part of uptake by the further addition of 10 mM L-phenylalanine. Concentration ratio = concentration in cell water/concentration in suspending medium. Incubation 1 min, 37°C. (Redrawn from data published by Christensen [30].)

this inhibitor. That the remaining component of transfer could still be inhibited is shown by the effect of the presence of 10 mM phenylalanine in addition to 50 mM α-(methylamino)isobutyrate; transfer of histidine was then almost completely abolished [30].

Whether non-competitive inhibitors can enable components of dual systems to be distinguished in this way is not known, but the effect of such inhibitors may vary with the type of substrate transferred or with the location of transfer [96, 106].

If an inhibitor affects both carriers of a dual system, analysis is more complex. For competitive inhibition of carrier transport there would be six constants, as shown by the composite equation which describes the general case (cf. equation (107)):

$$v_{\text{total}} = v_1 + v_2 = \underbrace{\frac{v_{\text{max}_1} \cdot [S]}{[S] + K_{m_1}\left(1 + \dfrac{[i]}{K_{i_1}}\right)}}_{\text{(Carrier 1)}} + \underbrace{\frac{v_{\text{max}_2} \cdot [S]}{[S] + K_{m_2}\left(1 + \dfrac{[i]}{K_{i_2}}\right)}}_{\text{(Carrier 2)}}. \qquad (113)$$

K_{i_1} and K_{i_2} are the values of inhibitor K_m for each of the two carriers.

This equation differs in no important way from equation (107), but it is algebraically more complex and may lead to certain consequences. For example, if the concentration of a competitive inhibitor is fixed, the characteristics of total transfer are those of a two-carrier system with K_{m_1} replaced by K_{a_1} (Apparent K_{m1}) and K_{m_2} by K_{a_2} (Apparent K_{m2}) (see Chapter 5). Alternatively, if the concentration of substrate is fixed but the concentration of inhibitor changed, the curve relating total rate of transfer to concentration of inhibitor will not be a rectangular hyperbola, as was the case with a one-carrier

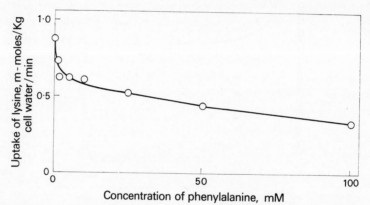

Fig. 49. Inhibition in a two-carrier system. The effect of different extracellular concentrations of the amino acid phenylalanine on the uptake of L-lysine by Ehrlich ascites carcinoma cells. Incubation 1 min, 37°C. The authors of the work suggest that two K_i values, one below 1 mM and one very much higher, would be required to account for the shape of the curve, assuming inhibition to be competitive. (Redrawn from data published by Christensen et al.[37].)

system (e.g. Figs. 33 and 39), but will be that of a quadratic function. This can be seen if in the above equation (eqn. 113) [i] is the only variable; the equation can then be simplified to the general algebraical form:

$$y = \frac{ax + b}{cx^2 + dx + e} \qquad (114)$$

where the symbol x corresponds to [i], and y to v_{total}.

A system which is probably of this type is shown in Fig. 49. The values plotted show the way in which the concentration of the amino acid phenylalanine may affect the entry of lysine into the Ehrlich ascites carcinoma cell. The experimental points plotted do not lie on a rectangular hyperbola but on a curve which could probably be described by the above equation (cf. Fig. 48). The authors of the paper from which the data were obtained claim that 'two K_i values, one below 1 mM and one very much higher, would be required to account for the shape of the curve' [37]. Unfortunately there are insufficient published data to know whether such a statement can be supported by experimental evidence. To be able to verify the statement it would be necessary to construct a hypothetical curve for comparison. To do this, two K_m values for lysine and for phenylalanine and two v_{max} values for lysine must be known, but at present only constants for phenylalanine appear to have been published (Table 4).

The presence of more than one carrier has sometimes been detected by comparing values measured as the K_i of an inhibitor with those measured as its K_m [16]. It will be remembered that K_i represents the K_m of an inhibitor when it is competing with a substrate for transport by a single common carrier (Chapter 5). If, however, the inhibitor but not the substrate is transported by a second carrier (Fig. 50, *left*), what is measured as the K_m for the inhibitor will be a pseudo-K_m (K_p) since it is a value which applies to two carriers at the same time and cannot be equated with K_i.

The same reasoning would apply where the substrate was transported by two carriers, but the inhibitor transported by one of them (Fig. 50, *right*). Since the inhibitor

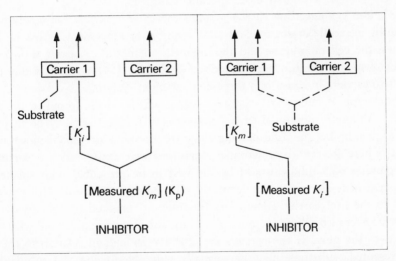

Fig. 50. How the measured value of K_i of a competitive inhibitor may differ from the measured value of its K_m. *Left:* the inhibitor is transported by two carriers (Carriers 1 and 2) but the substrate by only one (Carrier 1). K_i, which is measured from the effect of inhibitor on substrate, can only be related to Carrier 1, and therefore represents the K_m of the inhibitor for Carrier 1. When transport of the inhibitor on its own is measured, a single value measured as K_m (but in reality K_p) may be derived from its transport by both carriers (see text and Figs. 42–47). This value will differ from its K_m for Carrier 1 (i.e. K_i) unless its K_m is the same for both carriers. (If it is recognized that there are two carriers for the inhibitor, and the separate K_m for each obtained, the K_m for Carrier 1 should of course have the same value as K_i.) *Right:* the inhibitor is transported by one carrier (Carrier 1), but the substrate by two carriers (Carriers 1 and 2). The K_m of the inhibitor when transported on its own will be that for Carrier 1 only. The K_i which is measured will be derived from the result of inhibition of total transport of the substrate (i.e. transport by both carriers), and will therefore differ from the value of the K_m of the inhibitor.

will only affect part of the transport of the substrate, 'K_i' as measured would not be expected to have the same value as the K_m of the inhibitor.

Multi-carrier systems

So far, only two-carrier systems in parallel have been discussed. It is simple algebraically to extend carrier systems as a series as follows:

$$v_{\text{total}} = \frac{v_{\text{max}_1}\cdot[S]}{[S]+K_{m_1}} + \frac{v_{\text{max}_2}\cdot[S]}{[S]+K_{m_2}} + \frac{v_{\text{max}_3}\cdot[S]}{[S]+K_{m_3}} \cdots + \frac{v_{\text{max}_n}\cdot[S]}{[S]+K_{m_n}} \tag{115}$$

However, the analysis of experimental data which might suggest such a multi-carrier system becomes extremely hazardous, and, with the usual limit of resolution of such data, may become well-nigh impossible. We have already shown the difficulty that there may be in resolving only two components. As the number of additional components increases, the difficulty in distinguishing one from another becomes correspondingly greater, irrespective of experimental errors.

There is also evidence that carrier systems can exist in series. For a solute to penetrate into an intracellular organelle would require its movement across at least two membranes, the cell membrane and the organelle membrane. Carrier transport across both types of membrane has been described [102, 150, 154], so that in one cell there could be two carriers in series in a three-compartment system.

A further type of multi-carrier system is that which can be found in the tubule of the kidney. Although individual carrier systems [107, 140, 161] may be in parallel in the tubular membrane, the fluid passing along the tubule is presented in serial fashion. The kinetics here are even more complex than for carriers in parallel. A carrier system at the beginning of a tubule would be the first to be presented with substrate. The carrier system next along the tubule would then be presented with the concentration presented to the first carrier system less the amount removed by that carrier system. Each carrier system would thus remove part of the substrate available, and what remained would pass to the next. If the carriers are effective enough in relation to the concentration presented, substrate may be virtually eliminated from the tubule by serial extraction in this way.

In this chapter we have only treated multiple carrier systems from the point of view of one-way transport. To deal with it in terms of two-way or net transport becomes too complicated to discuss here, although in principle the theory is no more difficult than was shown for single-carrier transport in Chapter 3.

Chapter 7

Exchange, Exchange Diffusion and Countertransport

Although the movements of molecules in opposite directions across a membrane have often been viewed as mutually independent, they commonly appear to influence each other [69, 99, 100, 159]. Thus the entry of molecules into a cell can apparently increase the rate of exit of molecules of the same or a different species and vice versa. At first sight this might seem as if there were a direct linkage between the two movements, one movement appearing to 'drive' the other, but it can often be explained more simply in terms of purely physical effects or in terms of competition for sites on a carrier, as will be explained in more detail. Several terms, such as 'exchange', 'exchange diffusion' and 'countertransport' have been coined to distinguish between different types of observation, but all appear to be based on the same fundamental premises.

The concept of 'exchange' implies the replacement of one molecule by another and is identical in principle with competition (p. 56). It may be illustrated by the movement of ions across a selective cation-exchanger membrane [42, 147a]. It has been stated that if such a membrane is bathed on one side by a dilute solution of NaCl containing tracer Na⁺ and on the other by distilled water, only a relatively small amount of tracer Na⁺ will pass across; the addition of a small amount of NaCl to the distilled water will result in a great increase in the rate of movement of tracer Na⁺ across the membrane. In the first case a large proportion of the Na⁺ ions present is prevented from moving across by adsorption on to the membrane; in the second case some may be replaced or 'exchanged' by some of the Na⁺ ions added to the distilled water so that they are then free to move across. This type of exchange has been invoked to explain high fluxes of tracer ions in muscle [147a].

The simplest type of exchange across a membrane is 'exchange diffusion' (or, more accurately, 'exchange by diffusion'), which was originally defined in terms of a mole for mole exchange of molecules across an inert membrane to explain the replacement of unlabelled solute molecules by labelled molecules [84]. Let us imagine a membrane whose two surfaces are each in contact with a solution of molecules which can be transferred freely in either direction. As a result of thermal activity, molecules will move in random fashion in all directions. Some will move across the membrane from one side and may be replaced by molecules from the other side (see Chapter 2, p. 3), so that there is an exchange of molecules. This exchange will be one for one in unit time at equilibrium, and is also known as a mole for mole exchange. If only one molecular species is involved, exchange may not be detectable by simple chemical analysis, particularly when the total number of molecules in each part of the system is initially the same and remains unchanged, but the groups can usually be distinguished by the use of radioactive isotopes.

The rate of diffusion of a solute across a membrane may be influenced by the presence

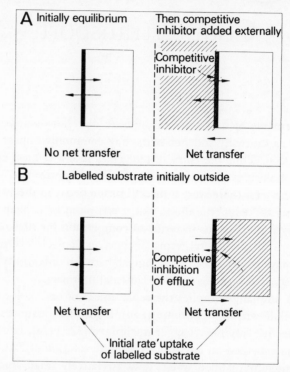

Fig. 51. Apparent stimulation of transfer of a substrate in one direction ostensibly by transfer of another substrate in the opposite direction. *Upper (A). Left:* an equilibrating transfer system with substrate in equilibrium across a cell membrane (*bold vertical line*), in which net transfer of substrate is zero, and the concentrations on either side of the membrane are the same. *Right:* a competitive inhibitor (*interrupted arrow*) has been introduced into the extracellular space. The rate of inward transfer of the substrate is consequently reduced, but the rate of outward transfer is initially almost unaltered. There is thus a net transfer outwards which immediately raises the extracellular concentration, so that outward transfer then proceeds against a concentration gradient. (As time passes the inhibitor may become distributed equally inside and outside the cell and act equally in both directions, so that when equilibrium is regained, the substrate moves once again at equal, although reduced, rates in each direction.) *Lower (B). Left:* 'initial rate' transfer of a labelled substrate into a cell is the difference between the rate of inward transfer of substrate from outside and the rate of outward transfer of substrate which has entered the cell. *Right:* unlabelled substrate has first been introduced into the cell. When labelled substrate enters, its outward transport is now reduced as a result of competition by the unlabelled substrate (represented by *interrupted arrow*). This results in an increase in the difference between inward and outward rates of transfer of substrate, which appears as an increase in 'initial rate' inwards.

of a second solute if their characteristics of diffusion differ. This was originally described for gases, but would equally well apply to solutes [59]. If one species of molecules, A, on one side of a membrane is smaller in size than a second species, B, on the other side, its movement through the membrane by random motion will be more rapid than that of B. Thus more A than B will pass through in unit time, and if the solution initially containing B is of finite volume a hydrostatic pressure (similar in its effect to osmotic pressure [12]) will be built up in the solution containing initially only B. The pressure will be 'relieved' by a compensating mass-flow of A and B together, that is, of the whole solution [59]. Thus the amount of B passing through the membrane in the presence of A will be greater than in its absence by an amount equal to the quantity contained in the mass-flow.

It is thus possible for movement of a solute across a membrane in one direction to influence the movement of another solute in the opposite direction without active participation by the membrane.

On the other hand, there have been a number of observations of the effect of transfer in one direction on transfer in the opposite direction which cannot be explained in such simple terms alone. Most can be explained more readily in terms of competition for carrier sites in a membrane during transfer. Although such observations are usually derived from 'initial rate' experiments, it must be remembered when interpreting them that it is the net result of two-way, not one-way, transfer which is being measured (see p. 38).

One example is the apparent ability of an equalizing transfer system to move substrate against a concentration gradient [115, 134, 158, 159]. If a sugar is allowed to come to equilibrium across the erythrocyte membrane (a membrane which incorporates an equalizing transfer system), the concentrations inside and outside the cell will be equal [122], and the rates of transfer in each direction will be the same (i.e. net transfer is zero) (Fig. 51, *upper left*). If a second, competing sugar is added to the external environment it reduces the rate of entry of the first sugar without initially affecting its outward rate (Fig. 51, *upper right*). A net outward transfer of the first sugar is then observed which, by thus increasing its external concentration, is in a direction against a concentration gradient. This effect will be transient for, as time passes, both sugars will come into equilibrium on either side of the membrane, although, as a result of competition, the one-way rates in opposite directions will be less than before, and will remain so. The magnitude of the initial effect should depend on the concentrations of the two sugars and on their relative abilities to occupy sites (i.e. on the K_m of each). Since competition is mutual, the net rate of transfer of the second sugar will also be affected; it may be decreased or increased depending on the kinetic parameters [136, 158].

The relationships can be expressed algebraically (using equation (86) and ignoring other influences such as altered osmotic pressure) thus:

$$V = v - v' = v_{max}\left(\underbrace{\frac{[S]}{[S] + K_m\left(1 + \frac{[i]}{K_i}\right)}}_{\text{(inward rate)}} - \underbrace{\frac{[S']}{[S'] + K_m\left(1 + \frac{[i']}{K_i}\right)}}_{\text{(outward rate)}}\right)$$

Here V is the net transport of the first sugar, which is represented by 'S'; 'i' represents the second sugar. Initially [S] (extracellular) = [S'] (intracellular), [i] is positive, and [i'] = 0, so that $v < v'$ and V is negative. [S] then becomes greater than [S']. As time passes, [i'] becomes positive, [i] gets less, v' decreases, v increases and eventually a new equilibrium is set up, where v and v' are both less than they were originally because of the presence of 'i'.

A similar phenomenon is the apparent increase in the rate of entry of a substrate into a cell as a result of the presence of substrate already in the cell. A tissue is 'loaded' with substrate and then immersed in a solution containing a labelled form of the same substrate; the labelled substrate may then appear to enter the tissue more rapidly than when the tissue had not been previously loaded [60, 69, 99, 110, 140]. This also can be explained in terms of competition. The measured rate of entry of labelled substrate is a balance between the quantities which enter and leave a cell during the experimental period (Fig. 51, *lower left*). Substrate initially in the cell competes with newly-entered substrate for outward carrier sites, so that less labelled substrate can move out again than would be the case if no substrate were in the cell (Fig. 51, *lower right*). The inward rate of transfer of labelled substrate is the same in both cases, so that the net rate of entry is greater when there is intracellular substrate already present than when there is not. It should follow that loading the cell with a different but related substrate which

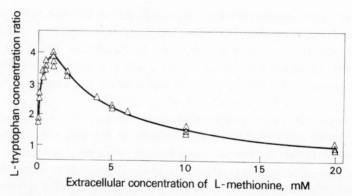

Fig. 52. Apparent stimulation of the uptake of one substrate in the presence of another (experimental example). The effect of various extracellular concentrations of the amino acid L-methionine on the uptake of L-tryptophan by Ehrlich ascites carcinoma cells. Values on the ordinate represent the ratio of the intracellular to the extracellular concentration of L-tryptophan. Initial extracellular concentration of L-tryptophan, 1 mM. Incubation 1 min. 37°C. At low concentrations extracellular methionine appeared to stimulate the uptake of tryptophan. This could be the result of a rapid entry of methionine into the cells at these concentrations (possibly partly by a route independent of tryptophan) and its subsequent inhibitory effect on the efflux of entered tryptophan (cf. Fig. 51, *lower*). The entry of tryptophan would also be interfered with by the extracellular methionine, but efflux must be affected more if an increase in the net uptake of tryptophan is to result. At high extracellular concentrations methionine would interfere primarily with the entry of tryptophan so that net uptake would be reduced. (Redrawn from data published by Jacquez [66].)

also competed for outward carrier sites would produce a similar effect, and this has been found to occur with a number of amino acids and sugars [69, 99, 140].

Yet another variation is the apparent increase in the rate of uptake of one compound by the presence of another, which is called 'competitive acceleration' [158] or 'competitive stimulation' [67] (to be distinguished from 'activation' (p. 69)). For example, the 'initial rate' of entry of the amino acid L-tryptophan into Ehrlich ascites carcinoma cells is increased by the presence of low extracellular concentrations of L-methionine but not by high concentrations (Fig. 52). This may be explained in similar terms to Fig. 51 (*lower*) if, after first entering the cell, methionine was able to occupy sites on an outward carrier more readily than tryptophan, perhaps because it was able to enter more rapidly, either by not competing so readily for the inward carrier or with the help of a second independent carrier; the latter seems to be the more likely explanation [67]. It would thus reduce the rate of exit of tryptophan more effectively than its rate of entry and so increase the intracellular concentration. At higher extracellular concentrations of methionine it would reduce the entry of tryptophan so much by direct inhibition of entry that overall intracellular accumulation of tryptophan was reduced. (In contrast, alanine produces only inhibition in these circumstances [66]; it must therefore be assumed that its effect on outward transfer was smaller than its effect on inward transfer.)

This general pattern of behaviour, in which a substrate transferred from one side of a membrane appears to affect the transfer of substrate in the opposite direction, has been termed both 'exchange diffusion' [29, 65, 69], and 'countertransport' [159]. Since 'exchange diffusion' refers to diffusion as such, the use of this term should rationally be confined to that type of movement, and not, as is often the case, to other types of exchange [25]. The term 'countertransport' should refer to carrier transport, as its meaning implies. One definition of it is 'an uphill transport which is independent of metabolic energy' [159], but this has perhaps too limited a meaning; it seems more sensible to apply 'countertransport' to all those cases in which a substrate transported by a carrier in one direction appears to influence transport in the opposite direction. In some cases it may not be possible to say with confidence whether a carrier is involved or not; when there is no evidence that it is, 'exchange diffusion' seems to be a reasonable description.

When experimental observations of this kind cannot be explained in terms of simple physical concepts such as mass flow, competition for carrier sites usually seems sufficient to explain them, although it is possible in some cases that competition alone may not be responsible [29]. In published work the subject may often seem to be made to appear more complicated than it is, but it is always preferable to interpret experimental findings in the simplest terms and not to invoke two mechanisms where one will do. As William of Ockham wrote: 'Essences are not to be multiplied without necessity' [112]. In this case the 'necessity' or 'demand' should only arise when there is a need to explain two groups of securely founded data which cannot be covered by a single interpretation.

Chapter 8

Some Experimental Aspects of the Kinetics of Transfer

Although the measurement of transfer is simple in principle, there are many different methods of investigation, each with its associated difficulties, its certainty of error, and the possibility of misinterpretation. The aim of this chapter is to show some of the difficulties which exist in standard procedures.

Movement of substrate into a cell

Transfer across cell membranes is commonly measured in terms of the uptake of substrate into a tissue. The ideal tissue would consist of separate identical cells, such as the erythrocyte [154, 159], or the Ehrlich ascites carcinoma cell [33], so that the external membranes across which transfer was taking place would be uniform and undiluted by other tissue components. The erythrocyte can be used to illustrate primarily equalizing transfer and the Ehrlich cell to show concentrative transfer, but this last needs serial cultures which may result in transport properties different from those of the original tissue [109]. Uptake into other tissues such as brain, intestinal epithelium and kidney appears to involve in each case several carrier systems [107] which may be a reflection of the variety of the constituent cell populations.

Uptake by isolated tissue is simpler to analyse than that into a tissue in an intact organism. In isolated tissue the concentration and total quantity of substrate presented to the tissue can be finely adjusted, but if substrate is injected into the bloodstream of an intact animal the required concentration is more difficult to arrange, and there may be by-pass shunts, change in blood flow, and inadequate mixing. On the other hand, of course, the data from experiments on isolated tissue are not directly comparable with those from tissue *in situ* (although they may be qualitatively similar); there may be components of metabolism which do not perform so well in the isolated state, although in some cases transfer may appear to be greatly enhanced [108].

An isolated tissue also has the advantage that the volume of fluid which surrounds it can be adjusted. This volume may be made small if the change in its substrate content is to be the prime measure of transfer, although this change may be non-linear with time. Alternatively, the volume may be made large if its substrate content is to remain fairly constant and the change in substrate content of the tissue is to be the prime measure of transfer, but here there may be errors associated with extraction of substrate from the tissue. In both cases there may also be sources of error, such as diversion to metabolism and intracellular adsorption which are common to various experimental techniques.

If substrate is being lost by metabolism the direction of error will depend on whether a change in intracellular or in extracellular concentration of substrate is the main indicator of transfer. If mainly tissue content is being measured, the concentration of substrate developed will be reduced in the presence of metabolic loss, which may suggest a relatively slow rate of transfer into the tissue. If the concentration of substrate in the surrounding

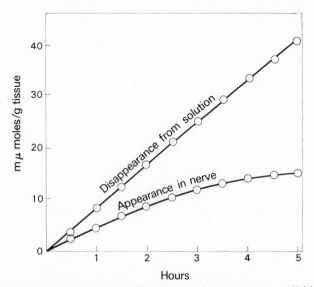

Fig. 53. Loss of substrate from one side of a membrane as an unreliable indicator of transfer performance. The amount of L-glutamate disappearing from a 1 mM solution bathing nerve tissue is compared with the amount appearing in the tissue after various times of incubation. The difference between the two curves reflects the amount of glutamate lost by metabolism. (Redrawn from data published by Wheeler & Boyarsky [152].)

fluid is the prime measure of transfer, disappearance from that fluid may be greater in the presence of metabolic loss, and may suggest a relatively high rate of transfer. This contrast is shown in Fig. 53 in which the metabolism of glutamate by nerve tissue results in a relatively high rate of disappearance from a suspending medium, but a relatively slow rate of appearance in the tissue [152]. It is thus clearly essential to measure substrate both inside and outside a tissue, and to determine whether there has been any loss of substrate from the system as a whole.

Movement of substrate out of a cell

The principles of the investigation of movement out of cells are the same as those for inward movement and may reveal similar kinetics, such as saturation (Fig. 54) and com-

H

petition [81]. The problems are mainly technical. Experiments are usually in two stages; the tissue is first 'loaded' with substrate taken up from the environment, and then the outward movement of substrate into a fresh environment, which may [110] or may not [81] contain substrate, is measured.

When investigating an equalizing transfer system, the tissue is 'loaded' by immersing it in a solution containing the concentration of substrate required and the system is then taken to near equilibrium, when the internal concentration approximates to the

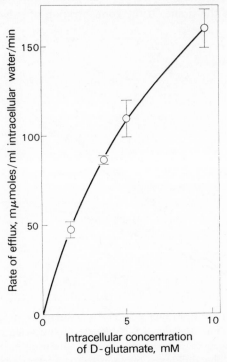

Fig. 54. Evidence for an outward carrier. Outward movement of D-glutamate from brain slices at various intracellular concentrations, showing progressive saturation with increase in concentration. Incubation 10 min, 37°C. Values shown are mean ±s.D. ($n = 4$). (Redrawn from data published by Levi *et al.* [81].)

external concentration. With a concentrative transfer system the internal concentration is not easily adjusted, since there is no simple relationship between the concentration inside and outside the tissue, and a system of trial and error may be necessary.

Outward transfer can then be measured, but the extracellular fluid must be changed frequently or the experimental period made short, since some of the substrate which has moved out will be transported inwards again, particularly if there is an extracellular space to impede the rapid dispersion of substrate molecules. Re-entry will also occur more readily in cells which can accumulate substrate against a concentration gradient, since inward transfer is then more 'effective' than outward transfer.

It is theoretically possible to determine kinetic constants for an outward carrier without measuring efflux or outward movement directly. If the constants for inward one-way transfer (K_m, v_{max} and K_D) are known, and the intracellular concentration ([S']) at equilibrium is determined, the appropriate values may be inserted into the equation (eqn. 39) representing equilibrium. By using a number of extracellular concentrations of substrate, the unknown values of the constants, K_{m_1} and v_{max_1}, for the outward carrier can then be calculated. The practical feasibility of this method is not known, but the size of error usually associated with constants for inward transfer will limit its accuracy.

Movement across whole tissue

The investigation of the movement of substrate across an entire tissue such as frog's skin or mammalian intestine *in vitro* is technically simple. The nature of the fluid on either side of the tissue can be easily controlled and the substrate whose movement is being observed can usually be measured without the need for extraction procedures. The volume and constituents of the fluid on either side of the tissue can be readily adjusted. The everted intestinal sac [162], which needs no special apparatus, is a simple and well-established preparation of this kind.

The use of whole tissue, however, suffers from the disadvantage that the kinetics of transfer may be complex [40]. Transfer across the wall of the intestine, for example, consists of movement across a number of cell boundaries, each with its own kinetic properties, so that total transfer may be a summation of several systems in series and in parallel. So-called kinetic constants obtained from such preparations may therefore be of a composite nature.

Determination of the concentration of a solute in cell water

After a tissue slice has been immersed in a suspending medium, the tissue for analysis will be contaminated by that medium (which may comprise up to 30 per cent. of the total fluid present [195]) and so the concentration of a solute in the cell fluid cannot be determined directly. For repeated studies of a tissue for which quantitative values are not required, the error resulting from the determination of the concentration of solute in the tissue mass as a whole may not be important, since one can usually assume it to be constant.

The true concentration of a solute in cell water can be determined by making a correction to its concentration in the water of the whole tissue sample. This consists of cell water, extracellular space water and adherent medium. Before discussing the correction factor, some of the terms must be defined (Fig. 55):

(i) *Total tissue fluid (TTF)* The total fluid in which the concentration of solute has been estimated, i.e. extracellular fluid + intracellular fluid + adherent suspending medium.

(ii) *Extracellular fluid (ECF)* All fluid outside the cells, i.e. fluid in the extracellular space + adherent suspending medium. The suspending medium is assumed to be in equilibrium with the fluid in the extracellular space. If it is assumed that inulin can

§H

penetrate the extracellular space but not the cell, the ECF would be equivalent to the total inulin space.

(iii) *Intracellular fluid (ICF)* Fluid which is contained within the cell membranes, and is assumed to be distributed uniformly through the cell.

The quantity or mass of solute (Q) in the intracellular fluid is equal to the quantity of solute in the total tissue fluid less the quantity of solute in the extracellular fluid, i.e.

$$Q_{ICF} = Q_{TTF} - Q_{ECF} \tag{116}$$

Now concentration of solute

$$= \frac{\text{quantity of solute}}{\text{volume of solvent}},$$

so that Quantity (Q)

$$= \text{concentration } ([S]) \times \text{volume (vol)} \tag{117}$$

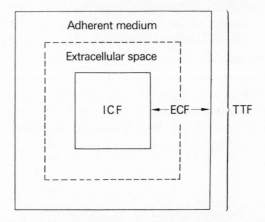

Fig. 55. Tissue spaces associated with the calculation of the intracellular concentration of solute. It is assumed that the adherent medium (in studies *in vitro*) and the fluid in the extracellular space are in equilibrium. ICF = intracellular fluid, ECF = extracellular fluid, TTF = total tissue fluid.

Hence:

$$([S]_{ICF} \times \text{vol}_{ICF}) = ([S]_{TTF} \times \text{vol}_{TTF}) - ([S]_{ECF} \times \text{vol}_{ECF}) \tag{118}$$

This can be rearranged:

$$[S]_{ICF} = \frac{([S]_{TTF} \times \text{vol}_{TTF}) - ([S]_{ECF} \times \text{vol}_{ECF})}{\text{vol}_{ICF}} \tag{119}$$

Since the volume of the intracellular fluid is the volume of the total tissue fluid less the volume of the extracellular fluid, i.e.

$$\text{vol}_{ICF} = \text{vol}_{TTF} - \text{vol}_{ECF}, \tag{120}$$

the equation can be rearranged thus:

$$[S]_{ICF} = \frac{([S]_{TTF} \times \text{vol}_{TTF}) - ([S]_{ECF} \times \text{vol}_{ECF})}{\text{vol}_{TTF} - \text{vol}_{ECF}}. \tag{121}$$

With tissue slices vol_{ECF} is taken to be the volume of the total inulin space and $[S]_{ECF}$ is the concentration of solute in the suspending medium, assumed to be in equilibrium with the extracellular space.

The method can be simplified if the inulin space and total fluid volume are assumed to have a constant relationship with each other in all samples. (This is an assumption which may not always be warranted [39].) Inulin space can be represented as a fraction (f) of the total fluid volume, which will itself then have a value of unity. The equation can now be simplified [83] to:

$$[S]_{ICF} = \frac{([S]_{TTF} \times 1) - ([S]_{suspending\ medium} \times f)}{1 - f}.$$ (122)

The accuracy of any value derived in this way depends of course on the validity of the basic assumptions. For example, the solute in the extracellular space is assumed to be in equilibrium with that in the suspending medium. If, however, it is, say, at a lower concentration the value used for $[S]_{ECF}$ will be too high, so that the calculated concentration of solute in intracellular fluid will be lower than the true concentration. An error of this sort could well exist in 'initial rate' experiments, particularly with a solute that was transported through the cell membrane rapidly and was therefore removed rapidly from the extracellular space.

Discrimination between adsorption and carrier transport

It is possible for a substrate adsorbed on to some structure, such as protein, in the cell to be extracted and measured as if it were in 'free' solution. Conversely, it is possible for substrate initially in 'free' solution in the cell to be adsorbed on to precipitated protein during extraction.

The kinetics of adsorption have been described earlier as an introduction to carrier transport (Chapter 3), and it should be clear that intracellular adsorption would show a kinetic pattern similar to that of carrier transport.

Methods in addition to simple kinetic analysis may therefore be needed to distinguish adsorption from carrier transport [145]. Some of these will be discussed and their merits criticized.

1. *Deproteinization*

Many solutes, particularly those which are singly ionized (in contrast to zwitterions) are commonly adsorbed on to protein. For example, antimony (as from antimony potassium tartrate) can be bound in (or on) liver cells in such a way as to give a total tissue concentration nearly forty times that in the environment [145].

It would seem logical that the precipitation of protein by a denaturing agent would cause the precipitation at the same time of the 'bound' solute, so that the solute which is 'free' in solution could be separated by filtration. Unfortunately, this may not be as simple a method for separating 'bound' from 'free' solute as it looks. First, all the protein may not be precipitated, so that some adsorbed solute may be measured as 'free' although attached to protein. Secondly, denaturation may increase the ability of protein to 'bind' solute so that more is removed from solution and the 'free' concentration reduced. This has been shown with liver slices, where, by applying heat (e.g. 80°C) or heavy metals (e.g. 2 mM $HgCl_2$) for 5 min to denature the protein, the adsorptive capacity has been shown to increase by about one-third [8], presumably by so altering the structure of the protein as to expose previously concealed adsorption sites. The precipitation of protein at low pH may similarly increase its ability to 'bind' solute.

2. *Disruption of cells*

The capacity of a cell to adsorb substrate may be measured by initially disrupting the cell (e.g. by homogenization) so that membrane carrier transport, but not adsorption, is eliminated.

This technique has been used for example, to determine the nature of the uptake of amines by choroid plexus [147]. With intact tissue, equilibration (from a 0·1 mM solution) was found to take about 2 h and resulted in a tissue/suspending medium concentration ratio of about 20 : 1. Iodoacetate, a metabolic inhibitor, reduced uptake by about 70 per cent. With homogenized tissue, equilibrium was apparently instantaneous; the particulate fraction/supernatant fraction concentration ratio was then about 5 : 1 and was unaffected by iodoacetate, suggesting that only adsorption was being measured, since iodoacetate has a marked inhibitory effect on transport [94]. (The two ratios obtained are, however, not directly comparable, since we do not know the adsorptive capacity at the concentration ruling intracellularly in the intact tissue.)

It is important to appreciate that uptake of solute by a homogenate is not in itself proof that adsorption is the mechanism responsible for high intracellular concentrations; subcellular particles surviving the homogenization may have their own 'carrier'-containing membranes [91, 119].

3. *Time taken to reach equilibrium*

Since attainment of the equilibrium of adsorption is generally more rapid than that of carrier transport, it might be thought that the rate of approach to equilibrium could be a way of distinguishing between the two, but this is not necessarily so.

Although adsorption by the contents of a ruptured cell may be almost instantaneous [147], adsorption by an intact cell may be slow [145], presumably owing to the delay in entry that may be caused by an intact cell membrane. Further, carrier transport also may be rapid, equilibrium being approached in some cases in less than a minute [24].

4. *Metabolic inhibitors*

It is usually assumed that adsorption is unaffected by metabolic inhibitors, and hence that adsorption is responsible for a high intracellular concentration of substrate which persists in the presence of an inhibitor of this sort [22, 145, 147]. However, this should not always be assumed to be the case. Firstly, metabolic inhibitors can vary in their effect on carrier transport [96, 106]. Secondly, a metabolic inhibitor might interfere with the structural integrity of the cell and hence of its individual components, such as protein responsible for adsorption Thirdly, physical forces, such as that of the Donnan equilibrium, can increase the intracellular concentration independently of carrier transport or adsorption.

Chapter 9

Conclusion

Experimental data may sometimes be interpreted in terms of a model, and this may suggest the direction of further experiment. An acceptable model must, of necessity, conform to experimental findings, and will be modified as experience accumulates; equations are a way of describing the model, and they also will be altered as the model changes. For example, we have seen that the kinetic model of a single carrier may be inadequate, and in certain cases a dual-carrier system has, apparently satisfactorily, replaced it. If there are reasons to suppose that data acquired later may conform to a particular model, those data must be accurate, sufficient in quantity, and viewed objectively in relation to the model if a reasonable comparison is to be made.

Earlier workers on carrier transfer [21, 153] were usually cautious and thorough; only when their most carefully observed findings could not be explained in terms of diffusion and a linear equation would they use non-linear forms which appeared to describe the data. Now that the concept of carrier transport is in many cases well established, the Michaelis–Menten equation (eqn. 27) is often applied uncritically in the presence of experimental data which may be insufficient to support its use.

It is not uncommon for published data to be inadequate or for assumptions to be made in such a way that the data are made to 'fit' a particular model. Figure 56, which has been taken (anonymously) from published work, is an example of this. It is a plot of $1/v$ against [i] at a fixed concentration of substrate (p. 78); the author did not state whether each point represents a single observation, but it would appear that this is so. If simple Michaelis–Menten kinetics applied, experimental points should lie on a straight line (see equation (100)). The most obvious relationship between the experimental points here, however, is a curve (interrupted line), yet the author appears to have believed in a straight-line relationship. The author has also given particular importance to the point representing $1/v$ when [i] = 0 by passing the drawn line through it. This value is the reciprocal of the rate of transfer of substrate in the absence of inhibitor, and there seems to be no reason to suppose that it should receive a 'weighting' not given to the other points. (The course of the actual 'best fit' straight line is shown by the two arrows, if, indeed, a straight line has to be drawn at all.)

As another example, we may cite instances in which values assigned to v_{max} and K_m have been obtained from experiments lasting up to 2 h, but employing kinetic equations

108

which describe one-way movement only [94, 131, 133, 161], and which should only be applied to 'initial rate' experiments. It has been claimed that the so-called K_m derived by graphical analysis of a 1 h experiment can have a value virtually the same as that of the K_m derived from a 5 min experiment [131]. The interest of such a claim (if it can be substantiated) is not only the experimental convenience which it may offer, but what it may have to say about the assumptions concerning the kinetics of either type of experiment. It seems that in certain circumstances only, such an identity of the two values may be possible without the need for major reconstruction of the simpler kinetics; if the equilibrium equation (eqn. 39) is appropriately rearranged, it can be shown algebraically that if inward transfer is by carrier, with or without diffusion, and outward

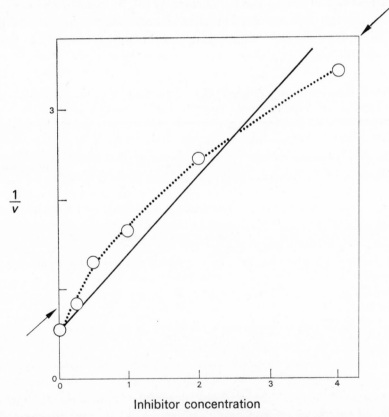

Fig. 56. Error of interpretation: straight-line graph of non-linear experimental values. If the reciprocal of the rate of transfer of substrate is plotted against the concentration of a competitive inhibitor (single-carrier kinetics), the experimental values should lie on a straight line (Chapter 5). In this instance, the solid line was drawn by the author of the original work, who has (i) assumed the existence of a straight-line relationship between the experimental values, and (ii) given excessive importance to the point at the lowest value of $1/v$. The straight line of best fit in fact lies between the two arrows, but even this would be unsatisfactory since the experimental points clearly lie on a curve, and to attribute to them a straight-line function would be misleading. Units are arbitrary. (Redrawn from published work; anonymous.)

transfer is by diffusion only (or by a mechanism indistinguishable from diffusion) a 'K_m' which agrees well with the 'initial rate' value can be derived from a resulting plot. The usefulness of this as a technique is limited by the prior need to ascertain the precise nature of individual components of transfer.

It is worth mentioning that in extended-time experiments, the net accumulation of substrate in tissue over the experimental period is sometimes expressed in terms of a 'rate', but this cannot have the same meaning as an 'instantaneous' or 'initial' rate (whose value is based on the assumption that transfer is uniform over a short period of time), since the increment in each small unit of time will change over the period of observation. To show the sort of inconsistency which may result from treating prolonged accumulation in this way, let us assume that effective equilibrium has been achieved after 1 h, so that the measured accumulation (i.e. the tissue content) will be virtually the same after 4 h as it is after 1 h. If it is then expressed in terms of a rate (mass of substrate accumulated per unit of time), the 4 h 'rate' would appear to be one-quarter of the 1 h 'rate', which is obviously absurd.

Table 8. Comparison of published values of Michaelis constants of amino acid transport in brain and spinal cord, derived from various sources as shown.

	K_m (mM)			
	Brain			Spinal cord
	Blasberg [16]	Smith [144]	Wheeler & Boyarsky [152]	Neal & Pickles [103]
Amino acid	Range of concentration used (mM)			
	0·1–2·0	0·1–10	0·001–0·1	0·001–0·1
D-alanine	2·4	1·47	—	—
L-alanine	1·0	0·44	—	—
ɣ-aminoisobutyrate	1·3	0·43	—	—
L-glutamate	0·48	—	0·035	—
Glycine	1·1	0·44	—	0·03

It is often difficult to assess the reliability of published kinetic constants, even when the experimental conditions are apparently satisfactory. As we have already said, statistical analysis of these constants is often absent from published work, and the published values for the constants relating to a particular substrate may vary considerably, as shown in Table 8. It is likely that moderate differences are due to small variations in experimental conditions (indeed a nearly two-fold difference has appeared in another instance to be related to the geographical location of the investigation [3]). Large differences, as can be seen in Table 8 for the two values attributed to K_m for glutamate, may, on the other hand, be due to limitation of the range of the concentrations used when applied to a transport system which does not follow simple Michaelis–Menten kinetics.

Each of these two values would then be the result of treating the data from each concentration range as if it represented single-carrier kinetics. If this was in reality a two-carrier system, one or both of these values would then have no real meaning in terms of a K_m (see Chapter 6).

The interpretation of apparent differences in the value of a kinetic constant of this sort is particularly important where different experimental conditions or different tissues are being investigated. For example, the published value for the K_m of the uptake of the amino acid glycine by spinal cord is markedly different from that for its uptake by brain (Table 8). This difference, also, might be due mainly to the difference in the ranges of concentration used in the investigations rather than to a property of the tissues.

It is in fact questionable whether any of the values in Table 8 really represent a K_m. They were all obtained over limited concentration ranges, and were presumably determined by means of single-carrier kinetic analysis, whereas there is evidence that all these amino acids may be transported in the brain by more than one carrier [16, 17, 18, 107]. It will hardly be necessary to point out that where the meaning or value of K_m is suspect, any value obtained for a K_i in similar circumstances must also be suspect.

It may now be clear that experimental design and the subsequent logical conduct of experiments are as important in the investigation of transfer kinetics as in other lines of enquiry. Given, say, four rates of uptake which happen to increase in some manner as substrate concentration increases, nothing seems more difficult to resist nor easier to do than to obtain a 'K_m' for the system, however little real meaning this may have. On the other hand, if data scrupulously acquired can be shown to fit Michaelis–Menten kinetics with a satisfactory probability, then to that extent it is reasonable to suppose that whatever may be causing the movement of substrate, it is behaving like the model. All models are partial; they behave in the 'mode' or manner of the object directly under observation, and the limit set to testing them depends on technical ability. The model described in these pages is a kinetic model, which should not be confused with a structural model, although we have used imaginary structures as an aid in describing it. It is a kinetic framework upon which the relationships of movement may be tested, and from which an attempt to predict behaviour may be made.

References

1 AFZELIUS B. (1964) *Anatomy of the Cell*. Chicago, University of Chicago Press.
2 AKEDO H. & CHRISTENSEN H.N. (1962) Nature of insulin action on amino acid uptake by the isolated diaphragm. *J. biol. Chem.* **237**, 118–22.
3 ALVARADO F. & CRANE R.K. (1964) Studies on the mechanism of intestinal absorption of sugars. VII. Phenylglycoside transport and its possible relationship to phlorizin inhibition of the active transport of sugars by the small intestine. *Biochim. biophys. Acta*, **93**, 116–35.
3a ANRAKU Y. (1967) The reduction and restoration of galactose transport in osmotically shocked cells of *Escherichia coli*. *J. biol. Chem.* **242**, 793–800.
4 ARIËNS E.J. (1964) *Molecular Pharmacology*, vol. 1. London, Academic Press.
5 ATKINSON M.R., JACKSON J.F. & MORTON R.K. (1961) Nicotinamide mononucleotide adenyltransferase of pig-liver nuclei. The effects of nicotinamide mononucleotide concentration and pH on dinucleotide synthesis. *Biochem. J.* **80**, 318–23.
6 AUGUSTINSSON K.-B. (1948) Cholinesterases. *Acta physiol. scand.* **15**, Suppl. 52, 1–182.
7 BAKER P.F. (1966) The sodium pump. *Endeavour*, **25**, 166–72.
8 BARBER-RILEY G. (1960) *Hepatic Adsorption of Bromsulphthalein*. M.D. thesis, University of Liverpool.
9 BARNETT J.E.G., JARVIS W.T.S. & MUNDAY K.A. (1968) Structural requirements for active intestinal sugar transport. The involvement of hydrogen bonds at C-1 and C-6 of the sugar. *Biochem. J.* **109**, 61–7.
10 BARRY R.J.C., EGGENTON J., SMYTH D.H. & WRIGHT E.M. (1967) Relation between sodium concentration, electrical potential and transfer capacity of rat small intestine. *J. Physiol.* **192**, 647–55.
11 BARRY B.A., MATTHEWS J. & SMYTH D.H. (1959) Water transfer in different parts of the rat intestine. *J. Physiol.* **149**, 78–79P.
12 BERKELEY EARL OF, & HARTLEY E.G.J. (1909) 'Dynamic' osmotic pressures. *Proc. R. Soc.* (*A*), **82**, 271–5.
13 BIDDER T.G. (1968) Hexose translocation across the blood-brain interface: configurational aspects. *J. Neurochem.* **15**, 867–74.
14 BIHLER I. (1969) Intestinal sugar transport: ionic activation and chemical specificity. *Biochim. biophys. Acta*, **183**, 169–81.
15 BLACKBURN K.J., FRENCH P.C. & MERRILLS R.J. (1967) 5-Hydroxytryptamine uptake by rat brain *in vitro*. *Life Sci.* **6**, 1653–63.
16 BLASBERG R. (1968) Specificity of cerebral amino acid transport: a kinetic analysis. In *Brain Barrier Systems* (*Progress in Brain Research*, vol. 29) (eds. Lajtha A. & Ford D.H.), pp. 245–56. Amsterdam, Elsevier.
17 BLASBERG R. & LAJTHA A. (1965) Substrate specificity of steady-state amino acid transport in mouse brain slices. *Archs Biochem. Biophys.* **112**, 361–77.
18 BLASBERG R. & LAJTHA A. (1966) Heterogeneity of the mediated transport systems of amino acid uptake in brain. *Brain Res.* **1**, 86–104.
19 BONTING S.L. (1970) Sodium-potassium activated adenosinetriphosphatase and cation transport. In *Membranes and Ion Transport* (ed. Bittar E.E.), vol. 1, pp. 257–363. London, Wiley-Interscience.

20 BORN G.V.R., DAY M. & STOCKBRIDGE A. (1967) The uptake of amines by human erythro-cytes *in vitro*. *J. Physiol.* **193**, 405–18.

21 BOWYER F. & WIDDAS W.F. (1956) The facilitated transfer of glucose and related compounds across the erythrocyte membrane. *Discuss. Faraday Soc.* **21**, 251–8.

22 BRAUER R.W. & PESSOTTI R.L. (1949) The removal of bromsulphthalein from blood plasma by the liver of the rat. *J. Pharmac. exp. Ther.* **97**, 358–70.

23 BRIGGS G.E. & HALDANE J.B.S. (1925) A note on the kinetics of enzyme action. *Biochem. J.* **19**, 338–9.

24 BRITTON H.G. (1964) Permeability of the human red cell to labelled glucose. *J. Physiol.* **170**, 1–20.

25 BRITTON H.G. (1970) Exchange diffusion. *Nature, Lond.* **225**, 746–7.

26 CARROLL L. (1872) *Through the Looking-glass, and What Alice Found There*, p. 100. London, Macmillan.

27 CHAPMAN D. (1970) The chemical and physical characteristics of biological membranes. In *Membranes and Ion Transport* (ed. Bittar E.E.), vol. 1, pp. 23–63. London, Wiley-Interscience.

28 CHRISTENSEN H.N. (1959) Active transport, with special reference to the amino acids. *Perspect. Biol. Med.* **2**, 228–42.

29 CHRISTENSEN H.N. (1962) *Biological Transport*. New York, W.A. Benjamin.

30 CHRISTENSEN H.N. (1968) Histidine transport into isolated animal cells. *Biochem. biophys. Acta*, **165**, 251–61.

31 CHRISTENSEN H.N. & HANDLOGTEN M.E. (1968) Modes of mediated exodus of amino acids from the Ehrlich ascites tumor cell. *J. biol. Chem.* **243**, 5428–38.

32 CHRISTENSEN H.N., HANDLOGTEN M.E., LAM I., TAGER H.S. & ZAND R. (1969) A bicyclic amino acid to improve discriminations among transport systems. *J. biol. Chem.* **244**, 1510–20.

33 CHRISTENSEN H.N. & LIANG M. (1965) An amino acid transport system of unassigned func-tion in the Ehrlich ascites tumor cell. *J. biol. Chem.* **240**, 3601–8.

34 CHRISTENSEN H.N. & LIANG M. (1966) Modes of uptake of benzylamine by the Ehrlich cell. *J. biol. Chem.* **241**, 5552–6.

35 CHRISTENSEN H.N. & LIANG M. (1966) Transport of diamino acids into the Ehrlich cell. *J. biol. Chem.* **241**, 5542–51.

36 CHRISTENSEN H.N., LIANG M. & ARCHER E.G. (1967) A distinct Na^+-requiring transport system for alanine, serine, cysteine, and similar amino acids. *J. biol. Chem.* **242**, 5237–46.

37 CHRISTENSEN H.N., OXENDER D.L., LIANG M. & VATZ K.A. (1965) The use of *N*-methyl-ation to direct the route of mediated transport of amino acids. *J. biol. Chem.* **240**, 3609–16.

38 CLAYMAN S. & SCHOLEFIELD P.G. (1969) The uptake of amino acids by mouse pancreas *in vitro*. IV. The role of exchange diffusion. *Biochim. biophys. Acta*, **173**, 277–89.

39 COHEN S.R., BLASBERG R., LEVI G. & LAJTHA A. (1968) Compartmentation of the inulin space in mouse brain slices. *J. Neurochem.* **15**, 707–20.

40 CRANE R.K. (1968) Absorption of sugars. In *Handbook of Physiology*, Section 6: *Alimentary Canal* (eds. Code C.F. & Heidel W.), vol. 3, pp. 1323–51. Washington, American Physiolo-gical Society.

41 CRANE R.K., FORSTNER G. & EICHHOLZ A. (1965) Studies on the mechanism of the intestinal absorption of sugars. X. An effect of Na^+ concentration on the apparent Michaelis constants for intestinal sugar transport *in vitro*. *Biochim. biophys. Acta*, **109**, 467–77.

42 CURRAN P.F. & SCHULTZ S.G. (1968) Transport across membranes: general principles. In *Handbook of Physiology*, Section 6: *Alimentary Canal* (eds. Code C.F. & Heidel W.), vol. 3, pp. 1217–43. Washington, American Physiological Society.

43 CURRAN P.F., SCHULTZ S.G., CHEZ R.A. & FUISZ R.E. (1967) Kinetic relations of the Na-amino acid interaction at the mucosal border of intestine. *J. gen. Physiol.* **50**, 1261–86.

44 DANIELLI J.F. (1954) Morphological and molecular aspects of active transport. *Symp. Soc. exp. Biol.* **8**, 502–16.

45 DAVSON H. (1970) *A Textbook of General Physiology*, 4th ed. London, J. & A. Churchill.

46 DAWSON R.M.C., ELLIOTT D.C., ELLIOTT W.H. & JONES K.M. (1959) *Data for Biochemical Research*, pp. 150–1. Oxford, Clarendon Press.

47 DIEDRICH D.F. (1966) Glucose transport carrier in dog kidney; its concentration and turn-over number. *Am. J. Physiol.* **211**, 581–7.

48 DIEDRICH D.F. & STRINGHAM C.H. (1970) Active site comparison of mutarotase with the glucose carrier in human erythrocytes. *Archs Biochem. Biophys.* **138**, 499–505.

49 DIXON M. (1953) The determination of enzyme inhibitor constants. *Biochem. J.* **55**, 170–1.

50 DIXON M. & WEBB E.C. (1964) *Enzymes*, 2nd ed. London, Longmans.

51 DOWD J.E. & RIGGS D.S. (1965) A comparison of estimates of Michaelis–Menten kinetic constants from various linear transformations. *J. biol. Chem.* **240**, 863–9.

52 EPSTEIN E. (1966) Dual pattern of ion absorption by plant cells and by plants. *Nature, Lond.* **212**, 1324–7.

53 EPSTEIN E. & RAINS D.W. (1965) Carrier-mediated cation transport in barley roots: kinetic evidence for a spectrum of active sites. *Proc. natn. Acad. Sci., U.S.A.* **53**, 1320–4.

54 EPSTEIN E., RAINS D.W. & ELZAM O.E. (1963) Resolution of dual mechanisms of potassium absorption by barley roots. *Proc. natn. Acad. Sci., U.S.A.* **49**, 684–92.

55 ERLANDER S.R. (1969) The structure of water. *Sci. J.* **5A**, 60–5.

55a FICK A. (1855) *Annln. Phys.* **94**, 59. Reported by ADAM N.K. (1962) *Physical Chemistry*, 3rd ed., p. 632. Oxford, Clarendon Press.

56 FINCH L.R. & HIRD F.J.R. (1960) The uptake of amino acids by isolated segments of rat intestine. II. A survey of affinity for uptake from rates of uptake and competition for uptake. *Biochim. biophys. Acta*, **43**, 278–87.

57 GLASSTONE S. (1948) *Textbook of Physical Chemistry*, 2nd ed. London, Macmillan.

58 HANES C.S. (1932) Studies of plant amylases. I. The effect of starch concentration upon the velocity of hydrolysis by the amylase of germinated barley. *Biochem. J.* **26**, 1406–21.

59 HARTLEY G.S. & CRANK J. (1949) Some fundamental definitions and concepts in diffusion processes. *Trans. Faraday Soc.* **45**, 801–18.

60 HEINZ E. & WALSH P.M. (1958) Exchange diffusion, transport, and intracellular level of amino acids in Ehrlich carcinoma cells. *J. biol. Chem.* **233**, 1488–93.

61 HILLMAN R.E. & ROSENBERG L.E. (1970) Amino acid transport by isolated mammalian renal tubules. III. Binding of L-proline by proximal tubule membranes. *Biochim. biophys. Acta*, **211**, 318–26.

62 HOFSTEE B.H.J. (1959) Non-inverted versus inverted plots in enzyme kinetics. *Nature, Lond.* **184**, 1296–8.

63 INUI Y. & CHRISTENSEN H.N. (1966) The Na^+-sensitive transport of neutral amino acids in the Ehrlich cell. *J. gen. Physiol.* **50**, 203–24.

64 IVERSEN L.L. & NEAL M.J. (1968) The uptake of [³H]GABA by slices of rat cerebral cortex. *J. Neurochem.* **15**, 1141–9.

65 JACQUEZ J.A. (1961) Transport and exchange diffusion of L-tryptophan in Ehrlich cells. *Am. J. Physiol.* **200**, 1063–8.

66 JACQUEZ J.A. (1963) Carrier-amino acid stoichiometry in amino acid transport in Ehrlich ascites cells. *Biochim. biophys. Acta*, **71**, 15–33.

67 JACQUEZ J.A. (1967) Competitive stimulation: further evidence for two carriers in the transport of neutral amino acids. *Biochim. biophys. Acta*, **135**, 751–5.

68 JOANNY P., CORRIOL J. & HILLMAN H. (1969) Uptake of monosaccharides by guinea-pig cerebral-cortex slices. *Biochem. J.* **112**, 367–71.

69 JOHNSTONE R.M. & SCHOLEFIELD P.G. (1965) Amino acid transport in tumor cells. *Adv. Cancer Res.* **9**, 143–226.

70 KLEINZELLER A., KOLÍNSKA J. & BENEŠ I. (1967) Transport of monosaccharides in kidney-cortex cells. *Biochem. J.* **104**, 852–60.

71 KLINGHOFFER K.A. (1935) Permeability of the red cell membrane to glucose. *Am. J. Physiol.* **111**, 231–42.

72 KOZAWA S. (1914) Beiträge zum arteigenen Verhalten der roten Blutkörperchen. III. Artidifferenzen in der Durchlässigkeit der roten Blutkörperchen. *Biochem. Z.* **60**, 231–56.

73 KUTTNER R., SIMS J.A. & GORDON M.W. (1961) The uptake of a metabolically inert amino acid by brain and other organs. *J. Neurochem.* **6**, 311–7.

74 LANGMUIR I. (1916) The constitution and fundamental properties of solids and liquids. Part I. Solids. *J. Am. chem. Soc.* **38**, 2221–95.

75 LEFEVRE P.G. (1948) Evidence of active transfer of certain non-electrolytes across the human red cell membrane. *J. gen. Physiol.* **31**, 505–27.

76 LEFEVRE P.G. (1954) The evidence for active transport of monosaccharides across the red cell membrane. *Symp. Soc. exp. Biol.* **8**, 118–35.

77 LEFEVRE P.G. (1962) Rate and affinity in human red blood cell sugar transport. *Am. J. Physiol.* **203**, 286–90.

78 LEFEVRE P.G. & DAVIES R.I. (1951) Active transport into the human erythrocyte: evidence from comparative kinetics and competition among monosaccharides. *J. gen. Physiol.* **34**, 515–24.

79 LEFEVRE P.G. & McGINNISS G.F. (1960) Tracer exchange *vs.* net uptake of glucose through human red cell surface. *J. gen. Physiol.* **44**, 87–103.

80 LEFEVRE P.G. & MARSHALL J.K. (1958) Conformational specificity in a biological sugar transport system. *Am. J. Physiol.* **194**, 333–7.

81 LEVI G., BLASBERG R. & LAJTHA A. (1966) Substrate specificity of cerebral amino acid exit *in vitro*. *Archs Biochem. Biophys.* **114**, 339–51.

82 LEVI G., CHERAYIL A. & LAJTHA A. (1965) Cerebral amino acid transport *in vitro*. III. Heterogeneity of exit. *J. Neurochem.* **12**, 757–70.

83 LEVI G., KANDERA J. & LAJTHA A. (1967) Control of cerebral metabolite levels. 1. Amino acid uptake and levels in various species. *Archs Biochem. Biophys.* **119**, 303–11.

84 LEVI H. & USSING H.H. (1948) The exchange of sodium and chloride ions across the fibre membrane of the isolated frog sartorius. *Acta physiol. scand.* **16**, 232–49.

85 LEVINE W., OXENDER D.L. & STEIN W.D. (1965) The substrate-facilitated transport of the glucose carrier across the human erythrocyte membrane. *Biochim. biophys. Acta*, **109**, 151–63.

86 LEVINE W. & STEIN W.D. (1966) The kinetic parameters of the monosaccharide transfer system of the human erythrocyte. *Biochim. biophys. Acta*, **127**, 179–93.

87 LINEWEAVER H. & BURK D. (1934) The determination of enzyme dissociation constants. *J. Am. chem. Soc.* **56**, 658–66.

88 LORENZO A.V. & CUTLER R.W.P. (1969) Amino acid transport by choroid plexus *in vitro*. *J. Neurochem.* **16**, 577–85.

89 LYON I. & CRANE R.K. (1966) Studies on transmural potentials *in vitro* in relation to intestinal absorption. I. Apparent Michaelis constants for Na^+-dependent sugar transport. *Biochim. biophys. Acta*, **112**, 278–91.

90 LUCY J.A. (1968) Theoretical and experimental models for biological membranes. In *Biological Membranes* (ed. Chapman D.), pp. 233–88. London, Academic Press.

91 MARCHBANKS R.M. (1968) The uptake of [^{14}C]choline into synaptosomes *in vitro*. *Biochem. J.* **110**, 533–41.

92 MARGOLIS R.K. & LAJTHA A. (1968) Ion dependence of amino acid uptake in brain slices. *Biochim. biophys. Acta*, **163**, 374–85.

93 MATTHEWS D.M., CRAFT I.L., GEDDES D.M., WISE I.J. & HYDE C.W. (1968) Absorption of glycine and glycine peptides from the small intestine of the rat. *Clin. Sci.* **35**, 415–24.

94 MATTHEWS D.M. & LASTER L. (1965) Kinetics of intestinal active transport of five neutral amino acids. *Am. J. Physiol.* **208**, 593–600.

95 MATTHEWS J. & SMYTH D.H. (1960) Two-stage transfer of glucose by the intestine. *J. Physiol.* **154**, 63–64P.

96 MATTHEWS J. & SMYTH D.H. (1961) The source of energy for fluid transfer by the intestine. *J. Physiol.* **158**, 13–14P.

97 MICHAELIS L. & MENTEN M.L. (1913) Die Kinetik der Invertinwirkung. *Biochem. Z.* **49** 333–69.

98 MILLER D.M. (1965) The kinetics of selective biological transport. I. Determination of transport constants for sugar movements in human erythrocytes. *Biophys. J.* **5**, 407–15.

99 MILLER D.M. (1968) The kinetics of selective biological transport. III. Erythrocyte-mono-saccharide transport data. *Biophys. J.* **8**, 1329–38.

100 MILLER D.M. (1968) The kinetics of selective biological transport. IV. Assessment of three carrier systems using the erythrocyte-monosaccharide transport data. *Biophys. J.* **8**, 1339–52.

101 NAFTALIN R.J. (1970) A model for sugar transport across red cell membranes without carriers. *Biochim. biophys. Acta,* **211**, 65–78.

102 NAVON S. & LAJTHA A. (1969) The uptake of amino acids by particulate fractions from brain. *Biochim. biophys. Acta,* **173**, 518–31.

103 NEAL M.J. & PICKLES H.G. (1969) Uptake of ^{14}C glycine by spinal cord. *Nature, Lond.* **222**, 679–80.

104 NEAME K.D. (1961) Uptake of amino acids by mouse brain slices. *J. Neurochem.* **6**, 358–66.

105 NEAME K.D. (1962) Uptake of L-histidine, L-proline, L-tyrosine and L-ornithine by brain, intestinal mucosa, testis, kidney, spleen, liver, heart muscle, skeletal muscle and erythrocytes of the rat *in vitro. J. Physiol.* **162**, 1–12.

106 NEAME K.D. (1964) Uptake of histidine, histamine and other imidazole derivatives by brain slices. *J. Neurochem.* **11**, 655–62.

107 NEAME K.D. (1968) A comparison of the transport systems for amino acids in brain, intestine, kidney and tumour. In *Brain Barrier Systems* (*Progress in Brain Research,* vol. 29) (eds. Lajtha A & Ford D.H.), pp. 185–96. Amsterdam, Elsevier.

108 NEAME K.D. (1968) Transport, metabolism and pharmacology of amino acids in brain. In *Applied Neurochemistry* (eds. Davison A.N. & Dobbing J.), pp. 119–77. Oxford, Blackwell Scientific Publications.

109 NEAME K.D. & GHADIALLY F.N. (1967) Uptake of L-histidine alone and in the presence of other amino acids by carcinogen-induced sarcomas of the rat *in vitro. Cancer Res.* **27**, 516–21.

110 NEAME K.D. & SMITH S.E. (1965) Uptake of D- and L-alanine by rat brain slices. *J. Neurochem.* **12**, 87–91.

111 NEWEY H. & SMYTH D.H. (1964) The transfer system for neutral amino acids in the rat small intestine. *J. Physiol.* **170**, 328–43.

112 OCKHAM, WILLIAM OF. Quoted by *Everyman's Encyclopaedia,* 4th ed. (1958), vol. 9 (ed. Bozman E.F.). London, Dent.

113 OXENDER D.L. (1965) Stereospecificity of amino acid transport for Ehrlich tumor cells. *J. biol. Chem.* **240**, 2976–82.

114 OXENDER D.L. & CHRISTENSEN H.N. (1963) Distinct mediating systems for the transport of neutral amino acids by the Ehrlich cell. *J. biol. Chem.* **238**, 3686–99.

114a PARDEE A.B., PRESTIDGE L.S., WHIPPLE M.B. & DREYFUSS J. (1966) A binding site for sulfate and its relation to sulfate transport into *Salmonella typhimurium. J. biol. Chem.* **241**, 3962–9.

114b PARDEE A.B. & WATANABE K. (1968) Location of sulfate-binding protein in *Salmonella typhimurium. J. Bact.* **96**, 1049–54.

115 PARK C.R., POST R.L., KALMAN C.F., WRIGHT J.H., JOHNSON L.H. & MORGAN H.E. (1956). The transport of glucose and other sugars across cell membranes and the effect of insulin. *Ciba Fdn. Colloq. Endocr.* **9**, 240–60.

116 PARKINSON C.N. (1958) *Parkinson's Law.* London, Murray.

117 PARSONS B.J. (1969) Binding of sugars to isolated brush borders. *Life Sci.* **8** (**II**), 939–42.

118 PERUTZ M.F., MUIRHEAD H., MAZZARELLA L., CROWTHER R.A., GREER J. & KILMARTIN J.V. (1969) Identification of residues responsible for the alkaline Bohr effect in haemoglobin. *Nature, Lond.* **222**, 1240–43.

119 PHILIPPU A., BURKAT U. & BECKE H. (1968) Uptake of norepinephrine by the isolated hypothalamic vesicles. *Life Sci.* **7** (I), 1009–17.

120 QUASTEL J.H. (1964) Symposium on transport reactions at the cell membrane. Introducorty survey. *Can. J. Biochem.* **42**, 907–16.

121 QUASTEL J.H. (1965) Molecular transport at cell membranes. *Proc. R. Soc. (B)*, **163**, 169–96.

122 REGEN D.M. & MORGAN H.E. (1964) Studies of the glucose-transport system in the rabbit erythrocyte. *Biochim. biophys. Acta*, **79**, 151–66.

123 REISER S. & CHRISTIANSEN P.A. (1967) Intestinal transport of valine as affected by ionic environment. *Am. J. Physiol.* **212**, 1297–302.

124 REISER S. & CHRISTIANSEN P.A. (1969) Intestinal transport of amino acids as affected by sugars. *Am. J. Physiol.* **216**, 915–24.

125 RICHARDS T.G. (1965) The plasma concentration of bromosulphalein (B.S.P.) after single intravenous injection in normal and abnormal human subjects. In *The Biliary System* (ed. Taylor W.), pp. 567–78. Oxford, Blackwell Scientific Publications.

126 RICHARDS T.G., SHORT A.H. & MIKULECKY M. (1968) Vyšetřování jaterní furcke pomoči bromsulfonftalein. *Bratisl. lék. Listy*, **50**, 585–96.

127 RICHARDS T.G., TINDALL V.R. & YOUNG A. (1959) A modification of the bromsulphthalein liver function test to predict the dye content of the liver and bile. *Clin. Sci.* **18**, 499–511.

128 RIGGS D.S. (1963) *The Mathematical Approach to Physiological Problems*. Baltimore, Williams & Wilkins.

129 ROBERTSON J.D. (1966) The unit membrane and the Danielli–Davson model. In *Intracellular Transport (Symposia of the International Society for Cell Biology*, vol. 5) (ed. Warren K.B.), pp. 1–31. New York, Academic Press.

130 ROBERTSON R.N. (1968) *Protons, Electrons, Phosphorylation and Active Transport*. Cambridge, University Press.

131 ROBINSON J.W.L. (1968) Interaction between neutral and dibasic amino acids for uptake by the rat intestine. *Eur. J. Biochem.* **7**, 78–89.

132 ROSENBERG L.E., ALBRECHT I. & SEGAL S. (1967) Lysine transport in human kidney: evidence for two systems. *Science, N.Y.* **155**, 1426–8.

133 ROSENBERG L.E., BLAIR A. & SEGAL S. (1961) Transport of amino acids by slices of rat-kidney cortex. *Biochim. biophys. Acta*, **54**, 479–88.

134 ROSENBERG T. & WILBRANDT W. (1957) Uphill transport induced by counterflow. *J. gen. Physiol.* **41**, 289–96.

135 ROSENBERG T. & WILBRANDT W. (1955) The kinetics of membrane transports involving chemical reactions. *Expl. Cell Res.* **9**, 49–67.

136 ROSENBERG T. & WILBRANDT W. (1963) Carrier transport uphill. I. General. *J. theoret. Biol.* **5**, 288–305.

137 ROTMAN B. & RADOJKOVIC J. (1964) Galactose transport in *Escherichia coli*. The mechanism underlying the retention of intracellular galactose. *J. biol. Chem.* **239**, 3153–6.

138 SCHOFFENIELS E. (1967) *Cellular Aspects of Membrane Permeability*. Oxford, Pergamon Press.

139 SCHOLEFIELD P.G. (1961) Competition between amino acids for transport into Ehrlich ascites carcinoma cells. *Can. J. Biochem. Physiol.* **39**, 1717–35.

139a SCHULTZ S.G. & CURRAN P.F. (1970) Coupled transport of sodium and organic solutes. *Physiol. Rev.* **50**, 637–718.

140 SCHWARTZMAN L., BLAIR A. & SEGAL S. (1967) Exchange diffusion of dibasic amino acids in rat-kidney cortex slices. *Biochim. biophys. Acta*, **135**, 120–6.

141 SEN A.K. & WIDDAS W.F. (1962) Determination of the temperature and pH dependence of glucose transfer across the human erythrocyte membrane measured by glucose exit. *J. Physiol.* **160**, 392–403.

142 SMITH H.W. (1951) *The Kidney*. New York, Oxford University Press.

143 SMITH S.E. (1963) Uptake of 5-hydroxy[^{14}C]tryptophan by rat and dog brain slices. *Br. J. Pharmac. Chemother*. **20**, 178–89.

144 SMITH S.E. (1967) Kinetics of neutral amino acid transport in rat brain *in vitro*. *J. Neurochem*. **14**, 291–300.

145 SMITH S.E. (1969) Uptake of antimony potassium tartrate by mouse liver slices. *Brit. J. Pharmac*. **37**, 476–84.

146 STEIN W.D. (1962) Spontaneous and enzyme-induced dimer formation and its role in membrane permeability. III. The mechanism of movement of glucose across the human erythrocyte membrane. *Biochim. biophys. Acta*, **59**, 66–77.

147 TOCHINO Y. & SCHANKER L.S. (1965) Transport of serotonin and norepinephrine by the rabbit choroid plexus *in vitro*. *Biochem. Pharmac*. **14**, 1557–66.

147a USSING H.H. (1947) Interpretation of the exchange of radio-sodium in isolated muscle. *Nature, Lond*. **160**, 262–3.

148 USSING H.H. (1952) Some aspects of the application of tracers in permeability studies. *Adv. Enzymol*. **13**, 21–65.

149 VAN BREEMEN D. & VAN BREEMEN C. (1969) Calcium exchange diffusion in a porous phospholipid ion-exchange membrane. *Nature, Lond*. **223**, 898–900.

150 VARON S. & WILBRANDT W. (1966) Na-dependent transport of γ-aminobutyric acid in subcellular brain particles. In *Intracellular Transport* (*Symposia of the International Society for Cell Biology*, vol. 5) (ed. Warren K.B.), pp. 119–39. New York, Academic Press.

151 WEST E.S. & TODD W.R. (1956) *Textbook of Biochemistry*, 2nd ed. New York, Macmillan.

152 WHEELER D.D. & BOYARSKY L.L. (1968) Influx of glutamic acid in peripheral nerve—characteristics of influx. *J. Neurochem*. **15**, 1019–31.

153 WIDDAS W.F. (1952) Inability of diffusion to account for placental glucose transfer in the sheep and consideration of the kinetics of a possible carrier transfer. *J. Physiol*. **118**, 23–39.

154 WIDDAS W.F. (1954) Facilitated transfer of hexoses across the human erythrocyte membrane. *J. Physiol*. **125**, 163–80.

155 WILBRANDT W. (1939) Die Permeabilität der roten Blutkörperchern für einfache Zucker. *Pfl˘gers Arch. ges. Physiol*. **241**, 302–9.

156 WILBRANDT W. (1956) The relation between rate and affinity in carrier transports. *J. cell. comp. Physiol*. **47**, 137–45.

157 WILBRANDT W. (1961) The sugar transport across the red cell membrane. In *Membrane Transport and Metabolism* (eds. Kleinzeller A. & Kotyk A.), pp. 388–97. London, Academic Press.

158 WILBRANDT W. (1969) Specific transport mechanisms in the erythrocyte membrane. *Experientia*, **25**, 673–7.

159 WILBRANDT W. & ROSENBERG T. (1961) The concept of carrier transport and its corollaries in pharmacology. *Pharmac. Rev*. **13**, 109–83.

160 WILKINSON G.N. (1961) Statistical estimations in enzyme kinetics. *Biochem. J*. **80**, 324–32.

161 WILSON O.H. & SCRIVER C.R. (1967) Specificity of transport of neutral and basic amino acids in rat kidney. *Am. J. Physiol*. **213**, 185–90.

162 WILSON T.H. & WISEMAN G. (1954) The use of sacs of everted small intestine for the study of the transference of substances from the mucosal to the serosal surface. *J. Physiol*. **123**, 116–25.

163 WINTER C.G. & CHRISTENSEN H.N. (1964) Migration of amino acids across the membrane of the human erythrocyte. *J. biol. Chem*. **239**, 872–8.

164 WISEMAN G. (1955) Preferential transference of amino acids from amino acid mixtures by sacs of everted small intestine of the golden hamster (*Mesocricetus auratus*). *J. Physiol*. **127**, 414–22.

165 WISEMAN G. (1968) Absorption of amino acids. In *Handbook of Physiology*, Section 6: *Alimentary Canal*, vol. 3 (eds. Code C.F. & Heidel W.), pp. 1277–307. Washington, American Physiological Society.

166 WYLD H.C. (1932) *The Universal Dictionary of the English Language*. London, Routledge.

Index